FREE

Free Study Tips DVD

In addition to the tips and content in this guide, we have created a FREE DVD with helpful study tips to further assist your exam preparation. **This FREE Study Tips DVD provides you with top-notch tips to conquer your exam and reach your goals.**

Our simple request in exchange for the strategy-packed DVD is that you email us your feedback about our study guide. We would love to hear what you thought about the guide, and we welcome any and all feedback—positive, negative, or neutral. It is our #1 goal to provide you with top-quality products and customer service.

To receive your **FREE Study Tips DVD**, email freedvd@apexprep.com. Please put "FREE DVD" in the subject line and put the following in the email:

a. The name of the study guide you purchased.

b. Your rating of the study guide on a scale of 1-5, with 5 being the highest score.

c. Any thoughts or feedback about your study guide.

d. Your first and last name and your mailing address, so we know where to send your free DVD!

Thank you!

Civil Service Exam Study Guide

Prep Book and Practice Test Questions with Detailed Answer Explanations [Firefighter, Police Officer, Clerical, etc.]

Matthew Lanni

Written and edited by APEX Publishing.

ISBN 13: 9781637754702
ISBN 10: 1637754701

For additional information or for bulk orders, contact info@apexprep.com

Table of Contents

Test Taking Strategies

1. Reading the Whole Question

A popular assumption in Western culture is the idea that we don't have enough time for anything. We speed while driving to work, we want to read an assignment for class as quickly as possible, or we want the line in the supermarket to dwindle faster. However, speeding through such events robs us from being able to thoroughly appreciate and understand what's happening around us. While taking a timed test, the feeling one might have while reading a question is to find the correct answer as quickly as possible. Although pace is important, don't let it deter you from reading the whole question. Test writers know how to subtly change a test question toward the end in various ways, such as adding a negative or changing focus. If the question has a passage, carefully read the whole passage as well before moving on to the questions. This will help you process the information in the passage rather than worrying about the questions you've just read and where to find them. A thorough understanding of the passage or question is an important way for test takers to be able to succeed on an exam.

2. Examining Every Answer Choice

Let's say we're at the market buying apples. The first apple we see on top of the heap may *look* like the best apple, but if we turn it over we can see bruising on the skin. We must examine several apples before deciding which apple is the best. Finding the correct answer choice is like finding the best apple. Although it's tempting to choose an answer that seems correct at first without reading the others, it's important to read each answer choice thoroughly before making a final decision on the answer. The aim of a test writer might be to get as close as possible to the correct answer, so watch out for subtle words that may indicate an answer is incorrect. Once the correct answer choice is selected, read the question again and the answer in response to make sure all your bases are covered.

3. Eliminating Wrong Answer Choices

Sometimes we become paralyzed when we are confronted with too many choices. Which frozen yogurt flavor is the tastiest? Which pair of shoes look the best with this outfit? What type of car will fill my needs as a consumer? If you are unsure of which answer would be the best to choose, it may help to use process of elimination. We use "filtering" all the time on sites such as eBay® or Craigslist® to eliminate the ads that are not right for us. We can do the same thing on an exam. Process of elimination is crossing out the answer choices we know for sure are wrong and leaving the ones that might be correct. It may help to cover up the incorrect answer choice. Covering incorrect choices is a psychological act that alleviates stress due to the brain being exposed to a smaller amount of information. Choosing between two answer choices is much easier than choosing between all of them, and you have a better chance of selecting the correct answer if you have less to focus on.

4. Sticking to the World of the Question

When we are attempting to answer questions, our minds will often wander away from the question and what it is asking. We begin to see answer choices that are true in the real world instead of true in the world of the question. It may be helpful to think of each test question as its own little world. This world may be different from ours. This world may know as a truth that the chicken came before the egg or may assert that two plus two equals five. Remember that, no matter what hypothetical nonsense may be in the question, assume it to be true. If the question states that the chicken came before the egg, then choose your answer based on that truth. Sticking to the world of the question means placing all of our biases and assumptions aside and relying on the question to guide us to the correct answer. If we are simply looking for answers that are correct based on our own judgment, then we may choose incorrectly. Remember an answer that is true does not necessarily answer the question.

5. Key Words

If you come across a complex test question that you have to read over and over again, try pulling out some key words from the question in order to understand what exactly it is asking. Key words may be words that surround the question, such as *main idea, analogous, parallel, resembles, structured,* or *defines.* The question may be asking for the main idea, or it may be asking you to define something. Deconstructing the sentence may also be helpful in making the question simpler before trying to answer it. This means taking the sentence apart and obtaining meaning in pieces, or separating the question from the foundation of the question. For example, let's look at this question:

> Given the author's description of the content of paleontology in the first paragraph, which of the following is most parallel to what it taught?

The question asks which one of the answers most *parallels* the following information: The *description* of paleontology in the first paragraph. The first step would be to see *how* paleontology is described in the first paragraph. Then, we would find an answer choice that parallels that description. The question seems complex at first, but after we deconstruct it, the answer becomes much more attainable.

6. Subtle Negatives

Negative words in question stems will be words such as *not, but, neither,* or *except.* Test writers often use these words in order to trick unsuspecting test takers into selecting the wrong answer—or, at least, to test their reading comprehension of the question. Many exams will feature the negative words in all caps (*which of the following is NOT an example*), but some questions will add the negative word seamlessly into the sentence. The following is an example of a subtle negative used in a question stem:

> According to the passage, which of the following is *not* considered to be an example of paleontology?

If we rush through the exam, we might skip that tiny word, *not,* inside the question, and choose an answer that is opposite of the correct choice. Again, it's important to read the question fully, and double check for any words that may negate the statement in any way.

7. Spotting the Hedges

The word "hedging" refers to language that remains vague or avoids absolute terminology. Absolute terminology consists of words like *always, never, all, every, just, only, none,* and *must.* Hedging refers to words like *seem, tend, might, most, some, sometimes, perhaps, possibly, probability,* and *often.* In some cases, we want to choose answer choices that use hedging and avoid answer choices that use absolute terminology. It's important to pay attention to what subject you are on and adjust your response accordingly.

8. Restating to Understand

Every now and then we come across questions that we don't understand. The language may be too complex, or the question is structured in a way that is meant to confuse the test taker. When you come across a question like this, it may be worth your time to rewrite or restate the question in your own words in order to understand it better. For example, let's look at the following complicated question:

> Which of the following words, if substituted for the word *parochial* in the first paragraph, would LEAST change the meaning of the sentence?

Let's restate the question in order to understand it better. We know that they want the word *parochial* replaced. We also know that this new word would "least" or "not" change the meaning of the sentence. Now let's try the sentence again:

> Which word could we replace with *parochial,* and it would not change the meaning?

Restating it this way, we see that the question is asking for a synonym. Now, let's restate the question so we can answer it better:

> Which word is a synonym for the word *parochial*?

Before we even look at the answer choices, we have a simpler, restated version of a complicated question.

9. Predicting the Answer

After you read the question, try predicting the answer *before* reading the answer choices. By formulating an answer in your mind, you will be less likely to be distracted by any wrong answer choices. Using predictions will also help you feel more confident in the answer choice you select. Once you've chosen your answer, go back and reread the question and answer choices to make sure you have the best fit. If you have no idea what the answer may be for a particular question, forego using this strategy.

10. Avoiding Patterns

One popular myth in grade school relating to standardized testing is that test writers will often put multiple-choice answers in patterns. A runoff example of this kind of thinking is that the most common answer choice is "C," with "B" following close behind. Or, some will advocate certain made-up word patterns that simply do not exist. Test writers do not arrange their correct answer choices in any kind of pattern; their choices are randomized. There may even be times where the correct answer choice will be the same letter for two or three questions in a row, but we have no way of knowing when or if this might happen. Instead of trying to figure out what choice the test writer probably set as being correct, focus on what the *best answer choice* would be out of the answers you are presented with. Use the tips above, general knowledge, and reading comprehension skills in order to best answer the question, rather than looking for patterns that do not exist.

FREE DVD OFFER

Achieving a high score on your exam depends not only on understanding the content, but also on understanding how to apply your knowledge and your command of test taking strategies. **Because your success is our primary goal, we offer a FREE Study Tips DVD. It provides top-notch test taking strategies to help you optimize your testing experience.**

Our simple request in exchange for the strategy-packed DVD is that you email us your feedback about our study guide.

To receive your **FREE Study Tips DVD**, email freedvd@apexprep.com. Please put "FREE DVD" in the subject line and put the following in the email:

a. The name of the study guide you purchased.

b. Your rating of the study guide on a scale of 1-5, with 5 being the highest score.

c. Any thoughts or feedback about your study guide.

d. Your first and last name and your mailing address, so we know where to send your free DVD!

Introduction

Function of the Test

The civil service in the United States recruits individuals into government positions that remain stable regardless of which political party is in power. Postal workers, law enforcement officers, and administrative assistants at federal agencies are just a few examples of civil servants. The federal civil service was initiated in 1871 under the Advisory Board of the Civil Service. The board was renamed the United States Civil Service Commission, and then it was divided into three distinct agencies: the Office of Personnel Management, the Merit Systems Protection Board, and the Federal Labor Relations Authority.

The civil service classifies appointees according to job purpose and grade. Civil servants take exams to advance from one grade to the next, and there is a six-month probationary period in each role. Pay is pegged to grade; the higher the grade level, the higher the pay. The most commonly known grade system is the General Schedule (GS). The GS consists of grades 1 through 15. Each grade also consists of steps from 1 to 10 (for example, GS-15, step 1).

Most civilians are part of the competitive service. This simply means that civilians must compete for federal positions and will be selected based on their qualifications. Part of this competition is the civil service exam.

Test Administration

The civil service exam is required of candidates who apply to work for state, local, or federal agencies, including the Secret Service, the U.S. Post Office, foreign affairs, and the Internal Revenue Service. Not all government job application processes will require a civil service exam, but many use it to narrow down a qualified pool of applicants. The exam is administered to potential firefighters, air traffic controllers, and federal accountants, among others.

There are different types of civil service exams, which are tailored to specific jobs. Whether you are seeking new employment in the civil service or trying to earn a promotion, there is a specific test you must take. However, every exam consists of the following sections: clerical skills, reasoning, reading and writing ability, and math. The specific administration may be online or on paper and will vary based on which entity is hiring.

Test Format

The civil service exam has written and multiple-choice sections. There may also be a verbal test, in which you will be asked questions pertaining to your position. If you are applying to be a firefighter or police officer, you will be required to take a physical fitness test.

The written portion of the exam is designed to test your abilities in reading comprehension, writing, and reasoning. You will be presented with questions that require you to write essays or short responses.

In the multiple-choice portion, some questions involve selecting the next number in a sequence or arranging content in numerical or alphabetical order. There are questions in which you will be asked to select the correct code for a set of data. Your math and logic skills will be tested in questions like these.

Scoring

To pass the civil service exam, you must achieve a score of 70% or higher. Any score below 70% is considered an automatic failure. Your grade will be given to you onsite or sent to you in email or regular mail. The hiring agency to which you applied is automatically notified of the results.

After the exam, the hiring agency ranks candidates from least to most eligible, based on the exam results. If you are deemed among the most eligible, you will receive a letter asking you to specify your interest in the position. If you indicate that you are interested, you will be selected for an interview. A successful interview will result in an offer and an opportunity to accept or decline the position.

Study Prep Plan

Breathe

Reducing stress is key when preparing for your test.

Build

Create a study plan to help you stay on track.

Begin

Stick with your study plan. You've got this!

1 Week Study Plan

Day 1	Day 2	Day 3	Day 4	Day 5	Day 6	Day 7
Verbal	Mathematics	Algebra	Geometry and Measurement	Data Analysis and Probability	Practice Test	Take your exam!

2 Week Study Plan

Day 1	Day 2	Day 3	Day 4	Day 5	Day 6	Day 7
Spelling	Reading Comprehension	Literal and Figurative Language	Number Operations	Rewriting Expressions Involving Radicals and Rational Exponents	Algebra	Solving Systems of Equations
Day 8	**Day 9**	**Day 10**	**Day 11**	**Day 12**	**Day 13**	**Day 14**
Translating Phrases and Sentences into Expressions, Equations, and Inequalities	Geometry and Measurement	Figure Nets	Data Analysis and Probability	Approximating the Probability of a Chance Event	Practice Test	Take your exam!

30 Day Study Plan

Day 1	Day 2	Day 3	Day 4	Day 5	Day 6	Day 7
Homophones	Analogies	Reading Comprehension	Narrative Writing	Analyzing Relationships within Passages	Function of Transitional Words and Phrases	Rational and Irrational Numbers

Day 8	Day 9	Day 10	Day 11	Day 12	Day 13	Day 14
Rewriting Expressions Involving Radicals and Rational Exponents	Reasoning Quantitatively and Using Units to Solve Problems	Solving Multi-Step Problems Involving Rational Numbers...	Interpreting Parts of an Expression	Writing Expressions in Equivalent Forms to Solve Problems	Solving Linear Equations in One Variable	Equivalent Expressions Involving Rational Exponents and Radicals

Day 15	Day 16	Day 17	Day 18	Day 19	Day 20	Day 21
Graphing Functions	Translating Phrases and Sentences into Expressions, Equations, and Inequalities	Writing a Function that Describes a Relationship Between Two Quantities	Transformations in the Plane	Properties of Polygons and Circles	The Pythagorean Theorem	Figure Nets

Day 22	Day 23	Day 24	Day 25	Day 26	Day 27	Day 28
Line Segments, Rays, and Lines	Summarizing Data Presented Verbally, Tabularly, and Graphically	Identifying the Line of Best Fit	Approximating the Probability of a Chance Event	Using Measures of Central Tendency to Draw Inferences About Populations	Using Statistics to Gain Information About a Population	Practice Test

Day 29	Day 30
Answer Explanations	Take your exam!

Verbal

Spelling

Homophones

Homophones are words with different meanings and spellings but the same pronunciation. These can be confusing for English Language Learners (ELLs) and beginning students, but even native English-speaking adults can find them problematic unless informed by context. Whereas listeners must rely entirely on context to *differentiate* between spoken homophone meanings, readers with good spelling knowledge have a distinct advantage since homophones are spelled differently. For instance, *their* is a possessive pronoun that means "belonging to them," while *there* indicates location, and *they're* is a contraction of *they are.* Despite different meanings, they all sound the same. *Lacks* is a present-tense, third-person singular verb that means "is deficient in," but *lax* is an adjective that means "loose, slack, relaxed, uncontrolled, or negligent." These two spellings, derivations, and meanings are completely different. With speech, listeners cannot know spelling and must use context, but with print, readers with can differentiate between homophones using spelling knowledge.

Homographs and Homonyms

Homographs are words that are spelled identically but have different meanings. If they also have different pronunciations, they are *heteronyms.* For instance, *tear* pronounced one way means a drop of liquid formed by the eye; pronounced another way, it means to rip. Homophones that are also homographs are called **homonyms**. For example, *bark* can mean the outside of a tree or a dog's vocalization; both meanings have the same spelling. *Stalk* can mean a plant stem or to pursue and/or harass somebody; these are spelled and pronounced the same. *Rose* can mean a flower or the past tense of *rise.*

Irregular Plurals

While many words in English can become plural by adding *–s* or *–es* to the end, there are some words that have irregular plural forms. One type includes words that are spelled the same whether they are singular or plural, such as *deer, fish, salmon, trout, sheep, moose, offspring, species, aircraft,* etc. Other irregular English plurals change form based on vowel shifts, linguistic mutations, or grammatical and spelling conventions from their languages of origin, like Latin or German. Some examples include *child* and *children; die* and *dice; foot* and *feet; goose* and *geese; louse* and *lice; man* and *men; mouse* and *mice; ox* and *oxen; person* and *people; tooth* and *teeth;* and *woman* and *women.*

Contractions

Contractions are formed by joining two words together, omitting one or more letters from one of the component words, and replacing the omitted letter(s) with an apostrophe. An obvious yet often forgotten rule for spelling contractions is to place the apostrophe where the letters were omitted. For example, *didn't* is a contraction of *did not*; therefore, the apostrophe replaces the "o" that is omitted from the "not." Another common error is confusing contractions with **possessives** because both include apostrophes, e.g., spelling the possessive *its* as "it's," which is a contraction of "it is"; spelling the possessive *their* as "they're," a contraction of "they are"; spelling the possessive *whose* as "who's," a contraction of "who is"; or spelling the possessive *your* as "you're," a contraction of "you are."

Frequently Misspelled Words

One source of spelling errors is not knowing whether to drop the final letter *e* from a word when its form is changed; some words retain the final *e* when another syllable is added while others lose it. For example, *true* becomes *truly, argue* becomes *arguing, come* becomes *coming, write* becomes *writing,* and *judge* becomes *judging.* In these examples, the final *e* is dropped before adding the ending. But *severe* becomes *severely, complete* becomes *completely, sincere* becomes *sincerely, argue* becomes *argued,* and *care* becomes *careful.* In these instances, the final *e* is retained before adding the ending. Note that some words, like *argue* in these examples, drops the final *e* when the *–ing* ending is added to indicate the participial form, but the regular past tense form keeps the *e* and adds a *–d* to make it *argued.*

Other commonly misspelled English words are those containing the vowel combinations *ei* and *ie.* Many people confuse these two. Some examples of words with the *ei* combination include:

> *ceiling, conceive, leisure, receive, weird, their, either, foreign, sovereign, neither, neighbors, seize, forfeit, counterfeit, height, weight, protein,* and *freight*

Words with *ie* include:

> *piece, believe, chief, field, friend, grief, relief, mischief, siege, niece, priest, fierce, pierce, achieve, retrieve, hygiene, science,* and *diesel*

A rule that also functions as a mnemonic device is, "I before E except after C, or when sounded like A as in 'neighbor' or 'weigh.'" However, it is obvious from the list above that many exceptions exist.

People often misspell certain words by confusing whether they have the vowel *a, e,* or *i.* For example, in the following correctly spelled words, the vowel in boldface is the one people typically get wrong by substituting one of the others for it:

> *cem**e**tery, quant**i**ties, ben**e**fit, priv**i**lege, unpleas**a**nt, sep**a**rate, independ**e**nt, excell**e**nt, cat**e**gories, indispens**a**ble,* and *irrelev**a**nt*

Another source of misspelling is whether or not to double consonants when adding suffixes. For example, double the last consonant before *–ed* and *–ing* endings in *controlled, beginning, forgetting, admitted, occurred, referred,* and *hopping;* but do not double the last consonant before the suffix in *shining, poured, sweating, loving, hating, smiling,* and *hoping.*

One final example of common misspellings involves either the failure to include silent letters or the converse of adding extraneous letters. If a letter is not pronounced in speech, it is easy to leave it out in writing. For example, some people omit the silent *u* in *g**u**arantee,* overlook the first *r* in *su**r**prise,* fail to double the *m* in *reco**mm**end,* and leave out the middle *i* from *aspirin.* The converse error, adding extra letters, is common in words like *until* by adding a second *l* at the end; or by inserting a superfluous syllabic *a* or *e* in the middle of *athletic,* reproducing a common mispronunciation.

Vocabulary

Context Clues, Syntax, and Structural Analysis in Unknown Words

When readers encounter an unfamiliar word in text, they can use the surrounding **context** to help determine the word's meaning. The text's overall subject matter, the specific chapter or section, and the immediate sentence context can all provide clues to help the reader understand the word. Among others, one category of context clues is grammar. For example, the position of a word in a sentence and its relationship to the other words can help the reader establish whether the unfamiliar word is a verb, a noun, an adjective, an adverb, etc. This narrows down the possible meanings of the word to one part of speech. However, this may be insufficient. In the sentence, "Many birds *migrate* twice yearly," the reader can determine that the italicized word is a verb. While it probably does not mean eat or drink (because birds would need to do those actions more than twice each year), it could mean travel, mate, lay eggs, hatch, molt, etc.

Another common context clue is a sentence that shows differences. Here's an example:

> Birds *incubate* their eggs outside of their bodies, unlike mammals.

Some readers may be unfamiliar with the word *incubate*. However, since the sentence includes the phrase "unlike mammals," the reader can infer that *incubate* relates to an aspect of prenatal development that mammals and birds do not have in common, such as the way they keep the embryo at a temperature suitable for development.

Readers can also determine word meanings from context clues based on logic. Here's an example:

> Birds are always looking out for predators that could attack their young.

The reader who is unfamiliar with the word *predator* could determine from the context of the sentence that predators usually prey upon baby birds and possibly other young animals. Readers might also use the context clue of etymology here, as *predator* and *prey* have the same root.

Analyzing Word Parts

Readers can learn some etymologies, or *origins*, of words and their parts, making it easier to break down new words into components and analyze their combined meanings. For example, the root word *soph* is Greek for "wise" or "knowledge." Knowing this informs the meanings of English words including *sophomore, sophisticated,* and *philosophy*. Those who also know that *phil* is Greek for "love" will realize that *philosophy* means "love of knowledge." They can then extend this knowledge of *phil* to understand *philanthropist* (one who loves people), *bibliophile* (book lover), *philharmonic* (loving harmony), *hydrophilic* (water-loving), and so on. In addition, *phob* derives from the Greek *phobos,* meaning "fear." Words with this root indicate fear of various things: *acrophobia* (fear of heights), *arachnophobia* (fear of spiders), *claustrophobia* (fear of enclosed spaces), *ergophobia* (fear of work), and *hydrophobia* (fear of water), among others.

Many words came into the early English language from sources like ancient Greek, Latin, and the Anglo-Saxon languages used by England's early tribes. By the Renaissance era, other influences included French, German, Italian, and Spanish. Today we can often discern English word meanings by knowing common roots and affixes, particularly from Greek and Latin.

The following is a list of common prefixes and their meanings:

Prefix	Definition	Examples
a-	without	atheist, agnostic
ad-	to, toward	advance
ante-	before	antecedent, antedate
anti-	opposing	antipathy, antidote

Prefix	Definition	Examples
auto-	self	autonomy, autobiography
bene-	well, good	benefit, benefactor
bi-	two	bisect, biennial
bio-	life	biology, biosphere
chron-	time	chronometer, synchronize
circum-	around	circumspect, circumference
com-	with, together	commotion, complicate
contra-	against, opposing	contradict, contravene
cred-	belief, trust	credible, credit
de-	from	depart
dem-	people	demographics, democracy
dis-	away, off, down, not	dissent, disappear
equi-	equal, equally	equivalent
ex-	from, out of	extract
for-	away, off, from	forget, forswear
fore-	before, previous	foretell, forefathers
homo-	same, equal	homogenized
hyper-	excessive, over	hypercritical, hypertension
in-	in, into	intrude, invade
inter-	among, between	intercede, interrupt
mal-	bad, poorly, not	malfunction
micr-	small	microbe, microscope
mis-	bad, poorly, not	misspell, misfire
mono-	one, single	monogamy, monologue
mor-	die, death	mortality, mortuary
neo-	new	neolithic, neoconservative
non-	not	nonentity, nonsense
omni-	all, everywhere	omniscient
over-	above	overbearing
pan-	all, entire	panorama, pandemonium
para-	beside, beyond	parallel, paradox
phil-	love, affection	philosophy, philanthropic
poly-	many	polymorphous, polygamous
pre-	before, previous	prevent, preclude
prim-	first, early	primitive, primary
pro-	forward, in place of	propel, pronoun
re-	back, backward, again	revoke, recur
sub-	under, beneath	subjugate, substitute
super-	above, extra	supersede, supernumerary
trans-	across, beyond, over	transact, transport
ultra-	beyond, excessively	ultramodern, ultrasonic, ultraviolet
un-	not, reverse of	unhappy, unlock
vis-	to see	visage, visible

The following is a list of common suffixes and their meanings:

Suffix	Definition	Examples
-able	likely, able to	capable, tolerable
-ance	act, condition	acceptance, vigilance
-ard	one that does excessively	drunkard, wizard
-ation	action, state	occupation, starvation
-cy	state, condition	accuracy, captaincy
-er	one who does	teacher
-esce	become, grow, continue	convalesce, acquiesce
esque	in the style of, like	picturesque, grotesque
-ess	feminine	waitress, lioness
-ful	full of, marked by	thankful, zestful
-ible	able, fit	edible, possible, divisible
-ion	action, result, state	union, fusion
-ish	suggesting, like	churlish, childish
-ism	act, manner, doctrine	barbarism, socialism
-ist	doer, believer	monopolist, socialist
-ition	action, result, state	sedition, expedition
-ity	quality, condition	acidity, civility
-ize	cause to be, treat with	sterilize, mechanize, criticize
-less	lacking, without	hopeless, countless
-like	like, similar	childlike, dreamlike
-ly	like, of the nature of	friendly, positively
-ment	means, result, action	refreshment, disappointment
-ness	quality, state	greatness, tallness
-or	doer, office, action	juror, elevator, honor
-ous	marked by, given to	religious, riotous
-some	apt to, showing	tiresome, lonesome
-th	act, state, quality	warmth, width
-ty	quality, state	enmity, activity

Identifying Different Interpretations

Language can function differently depending on its context. The same words can convey meaning in nuanced ways depending on the surrounding context and the style and tone of the composition. Just as how people can speak with a variety of tones and inflections that can alter the meaning of the same sentence, so too can writers insert tone into written words. Punctuation choice is one example of how the same sentence can be interpreted slightly differently.

Consider the following three sentences:

Camille hates dogs.

Camille hates dogs!

Camille hates dogs?

Although the wording is identical in these three simple sentences, the end punctuation affects the tone, and each option allows for a slightly different interpretation. The first sentence is simply stating that Camille hates dogs; the period at the end of the sentence elicits no significant emotional response. The exclamation point at the end of the second sentence could evoke surprise, exasperation, or urgency/alarm. This punctuation tends to evoke a feeling or surprise, exasperation, or urgency and alarm. Perhaps, for example, someone was about to introduce a large mastiff to Camille who had her back turned to the dog. One of Camille's friends who saw what was about to happen may have shouted that sentence in caution to prevent a terrified Camille. The last sentence is obviously a question, but that doesn't mean it's affectless. It could have been asked simply out of desire for clarification or confirmation, or it could have been asked in an incredulous or surprised tone because the speaker found it hard to believe that Camille could hate dogs. Other punctuation marks, especially commas, can also shape the way a sentence is read and interpreted.

The writer could further display tone by italicizing one of the words to indicate emphasis. Consider the difference between the following three examples:

> *Camille* hates dogs?
>
> Camille *hates* dogs?
>
> Camille hates *dogs*?

The first example places the emphasis on Camille. The speaker could be surprised that Camille, not another person, supposedly hates dogs. In the second example, the speaker's focus is on the word *hate.* He or she is seeking clarity or confirmation that Camille actually hates dogs (as opposed to simply being annoyed by them, allergic to them, afraid of them, etc. The italicized *dogs* in the last sentence indicates the speaker is verifying or expressing shock that Camille hates dogs in particular (rather than cats, spiders, rats, etc.).

In addition to punctuation and emphasis indicators, a word's connotation can also determine its interpretation. **Denotation**, a word's explicit definition, is often set in comparison to **connotation**, the emotional, cultural, social, or personal implications associated with a word. Denotation is more of an objective definition, whereas connotation can be more subjective, although many connotative meanings of words are similar for certain cultures. The denotative meanings of words are usually based on facts, and the connotative meanings of words are usually based on emotion.

Here are some examples of words and their denotative and connotative meanings in Western culture:

Word	Denotative Meaning	Connotative Meaning
Home	A permanent place where one lives, usually as a member of a family.	A place of warmth; a place of familiarity; comforting; a place of safety and security. "Home" usually has a positive connotation.
Snake	A long reptile with no limbs and strong jaws that moves along the ground; some snakes have a poisonous bite.	An evil omen; a slithery creature (human or nonhuman) that is deceitful or unwelcome. "Snake" usually has a negative connotation.
Winter	A season of the year that is the coldest, usually from December to February in the northern hemisphere and from June to August in the southern hemisphere.	Circle of life, especially that of death and dying; cold or icy; dark and gloomy; hibernation, sleep, or rest. Winter can have a negative connotation, although many who have access to heat may enjoy the snowy season from their homes.

How a Word's Context Affects Its Meaning

Some words can have a number of different meanings depending on how they are used. For example, the word *fly* has a different meaning in each of the following sentences:

- His trousers have a fly on them.
- He swatted the fly on his trousers.
- Those are some fly trousers.
- They went fly fishing.
- She hates to fly.
- If humans were meant to fly, they would have wings.

The context surrounding *fly* provides the reader with additional information needed to interpret the word's meaning.

Analogies

Verbal Analogies

Analogies are designed to test your knowledge of words and their relationships to one another. Making analogies is part of an important cognitive process that acts as the basis of metaphor and association. Analogical thinking is used in problem solving, creative thinking, argumentation, invention, communication, and memory, among other intellectual operations. Verbal analogies are a way to determine associations between objects, their signifying words, and each word's specific connotations. Many standardized exams include verbal analogies because of their usefulness in language and learning, especially in figurative language such as metaphors, similes, and allegories.

Analogy Format

An analogy will provide you with either two or three words to work with. Here is an example of a prompt with two words:

Cat is to **mammal** as

a. **shoe** is to **foot**.
b. **kitten** is to **canine**.
c. **dolphin** is to **amphibian**.
d. **lizard** is to **reptile.**
e. **lamp** is to **bedroom**.

Analogy questions require careful study of the relationship between the first pair of words. In this example, the first words have the relationship of **category** and **type.** A cat is a type of mammal; in other words, "cat" falls within the classification of "mammal." When determining the answer, you must find a pair in which the first word falls within the larger category of the second word. "Shoe" is not a type of foot, so Choice *A* is incorrect. "Canine" refers to dogs, not cats, so Choice *B* cannot be right. Choice *C* is incorrect because a dolphin is not an amphibian, but a mammal. "Lizard" is a type of reptile, so Choice *D* is correct. This pair of words has the same relationship as the original pair. Finally, Choice *E* is incorrect. A lamp may reside in a bedroom, and it could fall under the category of "bedroom furniture," but it is not a type of bedroom.

Now let's look at a prompt with three words:

Heat is to **scorching** as **cold** is to

a. burning.
b. freezing.
c. melting.
d. ice.
e. Alaska.

The first two words are related by **degree of intensity**. The first word is "heat," and "scorching" means a high degree of heat. To determine the answer, consider which word describes high intensity of cold. Choices *A* and *C*, burning and melting, are caused by heat, not cold, so these are incorrect. Choice *B*, "freezing," is an intensity of cold; Choice *D*, "ice," is an object that is really cold; Choice *E*, "Alaska," is a place that is really cold. While each of these choices is related to cold, it is important to look for which one forms the closest analogy to the word "scorching." The word is an adjective, not a noun, as ice and Alaska are. Therefore, "freezing," Choice *B*, is the best answer.

Types of Analogies

In its most basic form, an analogy compares two different things. An analogy is a pair of words that parallels the situation or relationship given in another pair of words. The table below give examples of the different types of analogies you may see:

Type of Analogy	Relationship	Example
Synonym	The words are alike in meaning.	**Happy** is to **joyous** as **sad** is to **somber**.
Antonym	The words are opposite in meaning.	**Lucky** is to **unfortunate** as **victorious** is to **defeated**.
Part to Whole	One word stands for a whole, and the other word stands as a part to that whole.	**Chapter** is to **novel** as **pupil** is to **eye**.
Category/Type	One thing belongs in a category of another thing.	**Screwdriver** is to **tool** as **apartment** is to **dwelling**.
Object to Function	A pair depicting a tool and the use of that tool.	**Shovel** is to **dig** as **oven** is to **bake**.
Degree of Intensity	The pair of words shows a difference in degree.	**Funny** is to **hysterical** as **interest** is to **adoration**.
Cause and Effect	The pair shows that one word is created by the other word.	**Hard work** is to **success** as **privilege** is to **comfort**.
Symbol and Representation	A word and its representation in the context of a culture.	**Rose** is to **love** as **flag** is to **patriotism**.
Performer to Related Action	A person and their related action.	**Professor** is to **teach** as **doctor** is to **heal**.

Reading Comprehension

Stated and Implied Ideas in a Text

A paragraph is made up of a series of related sentences that support a single main idea or message. The paragraph might clearly state the main idea or imply it through supporting details, guiding the reader to the desired conclusion.

If the main idea is not directly stated, the author must imply the main idea through the use of strong supporting details. These details could be comparisons of like ideas or contrasts of different ideas, factual evidence in graphs and statistics, quotes from experts, or vivid descriptions that evoke the desired emotional response.

Structure of a Text

Sequence structure is the order of events in which information is presented to the audience. Sometimes the text will be presented in chronological order, or sometimes it will be presented by displaying the most recent information first, then moving backwards in time. The sequence structure depends on the author, the context, and the audience. The sequence structure of a text also depends on the genre in which the text is written. Is it literary fiction? Is it a magazine article? Is it instructions for how to complete a certain task? Different genres will have different purposes for switching up the sequence.

The structure presented in literary fiction, called **narrative structure**, is the foundation on which the text moves. The narrative structure comes from the plot and setting. The plot is the sequence of events in the narrative that move the text forward through cause and effect. The setting is the place or time period in which the story takes place. Narrative structure has two main categories: linear and nonlinear.

Linear Narrative

A narrative is **linear** when it is told in chronological order. Traditional linear narratives will follow the plot diagram below depicting the narrative arc. The narrative arc consists of the exposition, conflict, rising action, climax, falling action, and resolution.

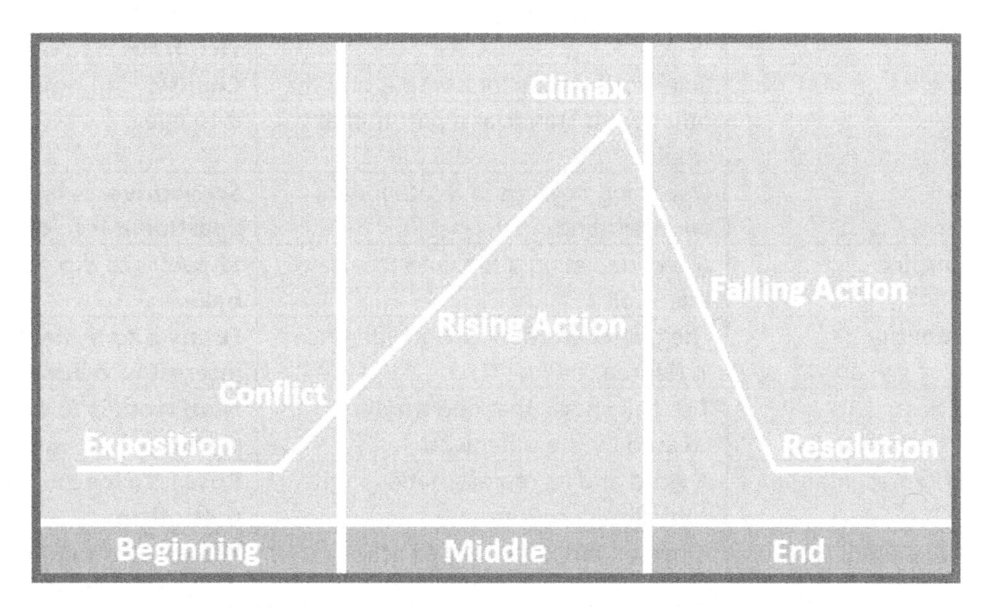

- **Exposition**: The exposition is in the beginning of a narrative and introduces the characters, setting, and background information of the story. The exposition provides the context for the upcoming narrative. Exposition literally means "a showing forth" in Latin.

- **Conflict**: In a traditional narrative, the conflict appears toward the beginning of the story after the audience becomes familiar with the characters and setting. The conflict is a single instance between characters, nature, or the self, in which the central character is forced to make a decision or move forward with some kind of action. The conflict presents something for the main character, or protagonist, to overcome.

- **Rising Action**: The rising action is the part of the story that leads into the climax. The rising action will develop the characters and plot while creating tension and suspense that eventually lead to the climax.

- **Climax**: The climax is the part of the story where the tension produced in the rising action comes to a culmination. The climax is the peak of the story. It is the height of the narrative, and it is usually either the most exciting part of the story or a turning point in the character's journey.

- **Falling Action**: The falling action happens as a result of the climax. Characters continue to develop, although there is a wrapping up of loose ends here. The falling action leads to the resolution.

- **Resolution**: The resolution is where the story comes to an end and usually leaves the reader with the satisfaction of knowing what happened within the story and why. However, stories do not always end in this fashion. Sometimes readers can be confused or frustrated at the end from lack of information or the absence of a happy ending.

Nonlinear Narrative

A **nonlinear** narrative deviates from the traditional narrative because it does not always follow the traditional plot structure of the narrative arc. Nonlinear narratives may include structures that are disjointed, circular, or disruptive, in the sense that they do not follow chronological order. ***In medias res*** is an example of a nonlinear structure. *In medias res* is Latin for "in the middle of things," which is how many ancient texts, especially epic poems, began their story, such as Homer's *Iliad*. Instead of having a clear exposition with a full development of characters, they would begin right in the middle of the action.

Many modernist texts in the late nineteenth and early twentieth centuries experimented with disjointed narratives, moving away from traditional linear narrative. Disjointed narratives are depicted in novels like *Catch 22*, where the author, Joseph Heller, structures the narrative based on free association of ideas rather than chronology. Another nonlinear narrative can be seen in the novel *Wuthering Heights*, written by Emily Brontë; after the first chapter, the narrative progresses retrospectively instead of chronologically. There seem to be two narratives in *Wuthering Heights* working at the same time: a present narrative as well as a past narrative. Authors employ disrupting narratives for various reasons; some use it for the purpose of creating situational irony for the readers, while some use it to create a certain effect, such as excitement, discomfort, or fear.

Sequence Structure in Technical Documents

The purpose of technical documents, such as instruction manuals, cookbooks, or "user-friendly" documents, is to provide information to users as clearly and efficiently as possible. In order to do this, the sequence structure in technical documents should be as straightforward as possible. This usually involves some kind of chronological order or a direct sequence of events. For example, someone who is reading an instruction manual on how to set up their Smart TV wants directions in a clear, simple, straightforward manner that does not confuse them or leave them guessing about the proper sequence.

Sequence Structure in Informational Texts

The structure of informational texts depends on the specific genre. For example, a newspaper article may start by stating an exciting event that happened, then talk about that event in chronological order. Many informational texts also use cause and effect structure, which describes an event and then identifies reasons for why that event occurred. Some essays may write about their subjects by way of comparison and contrast, which is a structure that compares two things or contrasts them to highlight their differences. Other documents, such as proposals, will have a problem to solution structure, where the document highlights some kind of problem and then offers a solution. Finally, some informational texts are written with lush details and description in order to captivate the audience, allowing them to visualize the information presented to them.

Arranging Ideas into an Outline

An **outline** is a system used to organize ideas. Outlines can be effective for readers and writers alike. Outlines and other graphic organizers can be helpful tools because they enable readers to organize important information logically. Outlines help readers to consider the author's main points and the supporting evidence for each main point. For fictional pieces, outlining can help readers keep track of the plot, characters, or chronology of the story. Usually, outlines start out with the main ideas and then branch out into subgroups or subsidiary thoughts or subjects. Not only do outlines provide a visual tool for readers to reflect on how events, ideas, evidence, or other key parts of the argument relate to one another, but they can also lead readers to improved comprehension and understanding.

The sample below demonstrates what a general outline looks like:

1. Introduction
    a. Background
    b. Thesis statement
2. Body
    a. Point A
        i. Supporting evidence
        ii. Supporting evidence
    b. Point B
        i. Supporting evidence
        ii. Supporting evidence
    c. Point C
        i. Supporting evidence
        ii. Supporting evidence
3. Conclusion
    a. Restatement of main points
    b. Memorable ending

Identifying the Author's Purpose in a Given Text

Authors may have many purposes for writing a specific text. They could be imparting information, entertaining their audience, expressing their own feelings, or trying to persuade their readers of a particular position. A single author may have one overriding purpose for writing or multiple reasons. An author may explicitly state their intention in the text, or the reader may need to infer that intention. When readers can identify the author's purpose, they are better able to analyze information in the text.

The following is a list of questions readers can ask in order to discern an author's purpose for writing a text:

- Did the author state their purpose for writing it?
- Does the title of the text give you any clues about its purpose?
- Was the purpose of the text to give information to readers?
- Did the author want to describe an event, issue, or individual?
- Was the text written to express emotions and thoughts?
- Did the author want to convince readers to consider a particular issue?
- Do you think the author's primary purpose was to entertain?
- Why do you think the author wrote this text from a certain point of view?
- What is your response to the text as a reader?

Rather than simply consuming the text, readers should attempt to interpret the information being presented. Being able to identify an author's purpose improves reading comprehension, develops critical thinking, and makes readers more likely to consider issues in depth before accepting the writer's viewpoints.

Authors seldom directly state their purposes in texts. Some readers may be confronted with nonfiction texts such as biographies, histories, magazine and newspaper articles, and instruction manuals, among others. To identify the purpose in nonfiction texts, students can ask the following questions:

- Is the author trying to teach something?
- Is the author trying to persuade the reader?
- Is the author imparting factual information only?
- Is this a reliable source?
- Does the author have some kind of hidden agenda?

To apply author purpose in nonfictional passages, students can also analyze sentence structure, word choice, and transitions to answer the aforementioned questions and to make inferences. For example, authors wanting to convince readers to view a topic negatively often choose words with negative connotations.

Narrative Writing

Narrative writing tells a story. The most prominent type of narrative writing is the fictional novel. Here are some examples:

- Mark Twain's *The Adventures of Tom Sawyer* and *The Adventures of Huckleberry Finn*
- Victor Hugo's *Les Misérables*
- Charles Dickens' *Great Expectations, David Copperfield*, and *A Tale of Two Cities*
- Jane Austen's *Northanger Abbey, Mansfield Park, Pride and Prejudice, Sense and Sensibility,* and *Emma*
- Toni Morrison's *Beloved, The Bluest Eye*, and *Song of Solomon*
- Gabriel García Márquez's *One Hundred Years of Solitude* and *Love in the Time of Cholera*

Nonfiction works can also appear in narrative form. Some authors may choose a narrative style to convey factual information about a topic because it engages readers in a different way. Many readers will find facts presented in the form of a story more entertaining and easier to follow.

Narrative writing tells a story, and the one telling the story is called the **narrator**. The narrator may be a fictional character telling the story from their own viewpoint. This narrator uses the first person (*I, me, my, mine* and *we, us, our,* and *ours*). The narrator may also be the author; for example, when Louisa May Alcott writes "Dear reader" in *Little Women*, she (the author) addresses us as readers. In this case, the novel is typically told in third person, referring to the characters as *he, she, they,* or *them*. Another more common technique is the omniscient narrator; in other words, the story is told by an unidentified individual who sees and knows everything about the events and characters—not only their externalized actions, but also their internalized feelings and thoughts. Second person narration, which addresses readers as *you* throughout the text, is more uncommon than the first and third person options.

Expository Writing

Expository writing is also known as informational writing. Unlike narrative writing, which tells a story, expository writing communicates and explains factual information. As such, the point of view of the author will necessarily be more objective. Expository writing does not appeal to emotions or reason, nor does it use subjective descriptions to sway the reader's opinion or thinking; rather, expository writing simply provides facts, evidence, observations, and objective descriptions of the subject matter. Some examples of expository writing include research reports, journal articles, books about history, academic textbooks, essays, how-to articles, user instruction manuals, news articles, and other factual journalistic reports.

Technical Writing

Technical writing is similar to expository writing because it provides factual and objective information. Indeed, it may even be considered a subcategory of expository writing. However, technical writing differs from expository writing in two ways: (1) it is specific to a particular field, discipline, or subject, and (2) it uses technical terminology that belongs only to that area. Writing that uses technical terms is intended only for an audience familiar with those terms. An example of technical writing would be a manual on computer programming and use.

Persuasive Writing

Persuasive or argumentative writing attempts to convince the reader to agree with the author's position. Some writers may respond to other writers' arguments by making reference to those authors or texts and then disagreeing with them. However, another common technique is for the author to anticipate opposing viewpoints, both from other authors and from readers. Writers persuade readers by appealing to the readers' reason and emotion, as well as to their own character and credibility. Aristotle called these appeals **logos**, **pathos**, and **ethos**, respectively.

Summarizing and Paraphrasing

An important skill is the ability to read a complex text and then reduce its length and complexity by focusing on the key events and details. A **summary** is a shortened version of the original text, written by the reader in their own words. The summary should be shorter than the original text, and it must include the most critical points.

In order to effectively summarize a complex text, it is necessary to understand the original source and identify the major points covered. It may be helpful to outline the original text to get the big picture and avoid getting bogged down in the minor details. For example, a summary wouldn't include a statistic from the original source unless it was the major focus of the text. It is also important for readers to use their own words but still retain the original meaning of the passage. The key to a good summary is emphasizing the main idea without changing the focus of the original information.

Complex texts will likely be more difficult to summarize. Readers must evaluate all points from the original source, filter out the unnecessary details, and maintain only the essential ideas. The summary often mirrors the original text's organizational structure. For example, in a problem-solution text structure, the author typically presents readers with a problem and then develops solutions through the course of the text. An effective summary would likely retain this general structure, rephrasing the problem and then reporting the most useful or plausible solutions.

Paraphrasing is somewhat similar to summarizing. It calls for the reader to take a small part of the passage and list or describe its main points. Paraphrasing is more than rewording the original passage, though. Like a summary, a paraphrase should be written in the reader's own words, while still retaining the meaning of the original source. The main difference between summarizing and paraphrasing is that a summary would be appropriate for a much larger text, while a paraphrase might focus on just a few lines of text. Effective paraphrasing will indicate an understanding of the original source, yet still help the reader expand on their interpretation. A paraphrase should neither add new information nor remove essential facts that change the meaning of the source.

Drawing Conclusions or Generalizations from a Text

Readers form conclusions about the main idea of a passage based on the evidence provided. As the reader gathers more details from the text, their conclusion could change or be proven by the evidence.

Generalizations are broad conclusions based on the way a reader interprets details. These generalizations might be correct, or the reader might have made inferences from the evidence that the author did not intend.

The reader should closely examine the evidence in the text to determine if the generalizations were actually implied by the author, or if they are incorrect.

For example:

1. Humans cannot live without water.
2. People use thousands of gallons of water a year to keep their lawns green.

Based on these details, the reader concludes that *people who water their lawns are wasting water.*

The author has not offered evidence that watering lawns is wasteful, so the generalization is not supported. If the passage had included the detail that *scientists say water will soon be dangerously scarce*, then the generalization would be supported. On the other hand, the following statement would contradict the reader's generalization: *Lawns also need water to grow and sustain life.*

Making inferences and drawing conclusions both require readers to fill in information the writer has omitted. To make an inference or draw a conclusion about the text, test takers should observe all facts and arguments the writer has presented. Consider the following example:

Nutritionist: "Many bodybuilders turn to whey protein as a way to repair their muscle tissue after working out. Recent studies are showing that using whey as a source of protein is linked to prostate cancer in men. Bodybuilders who use whey protein may consider switching to a plant-based protein source in order to avoid the negative effects that could come with whey protein consumption."

Which of the following most accurately expresses the conclusion of the nutritionist's argument?

a. Whey protein is an excellent way to repair muscles after a workout.
b. Bodybuilders should switch from whey to plant-based protein.
c. Whey protein causes every single instance of prostate cancer in men.
d. We still do not know the causes of prostate cancer in men.
e. Whey protein is not effective at repairing muscle tissue after working out.

The correct answer choice is *B*: bodybuilders should switch from whey to plant-based protein. We can gather this from the entirety of the passage, as it begins with what kind of protein bodybuilders consume, the dangers of that protein, and what kind of protein to switch to. Choice *A* is incorrect; this is the opposite of what the passage states. When reading through answer choices, it is important to look for choices that include the words "every," "always," or "all." In many instances, absolute answer choices will not be the correct answer. Take Choice *C* for example: the passage does not state that whey protein causes "every single instance" of prostate cancer in men, only that it is *linked* to prostate cancer in men. Choice *D* is incorrect; although the nutritionist does not list all causes of prostate cancer in men, he/she does not conclude that we do not know any causes of prostate cancer in men. Finally, Choice *E* is incorrect because the argument does not provide any evidence that whey protein is not effective at repairing muscle tissue; rather, it raises concerns about the potentially dangerous side effects.

The key to drawing well-supported conclusions is to read the question a few times and then paraphrase the passage. This will help you get an idea of the passage's conclusion before reading the different answer choices. Remember that drawing a conclusion is different than making an assumption. Conclusions must rely solely on the facts of the passage. Making an assumption goes beyond the facts of the passage, so be careful of answer choices depicting assumptions instead of passage-based conclusions.

Making Inferences Based on Information from a Text

An **inference** is an educated guess or conclusion based on sound evidence and reasoning within the text. The test may include multiple-choice questions asking about the logical conclusion that can be drawn from reading a text, and you will have to identify the choice that unavoidably leads to that conclusion.

Here is an example:

> Fred purchased the newest PC available on the market. Therefore, he purchased the most expensive PC in the computer store.
>
> What can one assume for this conclusion to follow logically?
>
> a. Fred enjoys purchasing expensive items.
> b. PCs are some of the most expensive personal technology products available.
> c. The newest PC is the most expensive one.

The premise of the text is the first sentence: *Fred purchased the newest PC*. The conclusion is the second sentence: *Fred purchased the most expensive PC*. Recent release and price are two different factors; the difference between them is the logical gap. To eliminate the gap, one must connect the new information from the conclusion with the pertinent information from the premise. In this example, there must be a connection between product recency and product price. Therefore, a possible bridge to the logical gap could be a sentence stating that the newest PCs always cost the most.

Analyzing Relationships within Passages

In a passage, the relationships between people, events, and ideas may be clearly stated, or the reader might have to infer them based on textual clues. Again, to *infer* means to arrive at a conclusion based on evidence, clues, or facts.

People might be related through connections like family or friendship, or through events that link them directly or indirectly. In the passage, relationships may be described in background information or dialogue, or they may be implied through interactions between the characters.

Events and ideas in a passage can be related through sequence, comparison, or cause and effect. Relating events and ideas through sequence means ordering them chronologically, alphabetically, or geographically. An author can also relate ideas and events through comparison, showing the similarities and differences. Finally, cause and effect displays a relationship where one event or idea triggers another.

Sequence
When ideas are related through sequence, as in a series of events, the author typically uses signal words like *after, and then, while,* and *before.*

Compare and Contrast
Ideas are often connected through comparisons of their similarities using the signal words such as *like* or *and.* Ideas can also be connected through a contrast of their differences using signal words such as *but* or *however.*

Cause and Effect
In cause and effect relationships, the cause is an event or circumstance that occurs before and is directly responsible for the effect. Signal words such as *because, due to,* and *as a result of* indicate a cause and effect relationship. When the relationship is implied, the reader must use clues in the passage to infer that the effect resulted from the cause.

Literal and Figurative Language

When the meaning of a text requires directness, the author will use literal language to provide clarity. On the other hand, an author will use figurative language to produce an emotional effect or facilitate a deeper understanding of a word or passage. For example, a set of instructions on how to use a computer would require literal language. However, a commentary on the social implications of immigration bans might contain a wide range of figurative language to elicit an empathetic response. A single text can have a mixture of both literal and figurative language.

It is important to be able to recognize and interpret figurative, or non-literal, language. **Literal** statements rely directly on the denotations of words and express exactly what is happening in reality. **Figurative language** uses non-literal expressions to present information in a creative way. Consider the following sentences:

a. His pillow was very soft, and he fell asleep quickly.

b. His pillow was a fluffy cloud, and he floated away on it to the dream world.

Sentence *A* is literal, employing only the real meanings of each word. Sentence *B* is figurative. It uses a metaphor by stating that his pillow was a cloud. Of course, he is not actually sleeping on a cloud, but the reader can draw on images of clouds as light, soft, fluffy, and relaxing to get a sense of how the character felt as he fell asleep. Also, in sentence *B*, the pillow becomes a vehicle that transports him to a magical dream world. The character is not literally floating through the air—he is simply falling asleep! By utilizing figurative language, the author creates a scene of peace, comfort, and relaxation that conveys stronger emotions and more creative imagery than the purely literal sentence. Figurative language is not meant to be taken literally, but it is useful when the author wants to produce an emotional effect in the reader or add heightened complexity to the text's meaning.

Figurative language is used more heavily in literary fiction, poetry, critical theory, and speeches. It goes beyond literal language, allowing readers to form associations they wouldn't normally form. Using language in a figurative sense appeals to the reader's imagination. It is important to remember that words signify objects and ideas and are not the objects and ideas themselves. Figurative language can highlight this detachment by creating multiple associations, but also points to the fact that language is fluid and capable of creating a world full of linguistic possibilities. It can be argued that figurative language is the heart of communication even outside of fiction and poetry. People connect through humor, metaphors, cultural allusions, puns, and symbolism in their everyday rhetoric.

Similes and Metaphors

A **simile** is a comparison of two things using *like*, *than*, or *as*. A simile usually takes objects that have no apparent connection, such as a mind and an orchid, and compares them:

> His mind was as complex and rare as a field of ghost orchids.

Similes encourage new, fresh perspectives on objects or ideas that would not otherwise occur. Unlike similes, **metaphors** are comparisons that do not use *like*, *than*, or *as*. Metaphors compare objects or ideas directly, asserting that something *is* a certain thing, even if it isn't. A metaphor from the above example would be:

> His mind was a field of ghost orchids.

Thus, similes highlight the comparison by focusing on the figurative side of the language, elucidating the author's intent. Metaphors, however, provide a beautiful yet somewhat equivocal comparison. The following is another example of a metaphor used by the author Virginia Woolf:

> Books are the mirrors of the soul.

Metaphors consist of two parts: a tenor and a vehicle. Tenor refers to the object being described, and vehicle refers to the figurative language making the comparison. In this example, the tenor is "books," and the vehicle is "mirrors of the soul." Perhaps the author meant to say that written language (books) reflects a person's most inner thoughts and desires.

Dead metaphors are phrases that have been overused to the point where the figurative language has taken on a literal meaning, like "crystal clear." This phrase is in such popular use that the meaning seems literal ("perfectly clear") even when it is not.

Finally, an **extended metaphor** is one that goes on for several paragraphs, or even an entire text. "On First Looking into Chapman's Homer," a poem by John Keats, begins, "Much have I travell'd in the realms of gold," and goes on to explain the first time he hears Chapman's translation of Homer's writing. The extended metaphor begins in the first line as Keats compares travelling into "realms of gold" to the act of hearing a certain kind of literature for the first time. The metaphor continues through the end of the poem where Keats stands "Silent, upon a peak in Darien," having heard the end of Chapman's translation. Keats gained insight into new lands (new text) and was richer because of it.

Other Types of Figurative Language

The following are brief definitions and examples of popular figurative language:

- **Onomatopoeia**: A word that, when spoken, imitates the sound to which it refers. Example: "We heard a loud *boom* while driving to the beach yesterday."
- **Personification**: When human characteristics are given to animals, inanimate objects, or abstractions. An example would be in William Wordsworth's poem "Daffodils" where he sees a "crowd . . . / of golden daffodils . . . / Fluttering and dancing in the breeze." Dancing is usually a characteristic attributed solely to humans, but Wordsworth personifies the daffodils here as a crowd of people dancing.
- **Juxtaposition**: Juxtaposition places two objects side by side for comparison or contrast. For example, Milton juxtaposes God and Satan in "Paradise Lost."
- **Paradox**: A paradox is a statement that appears self-contradictory but is actually true. One example of a paradox is when Socrates said, "I know one thing; that I know nothing." Seemingly, if Socrates knew nothing, he would not know that he knew nothing. However, he is using figurative language not to say that he literally knows nothing, but that true wisdom begins with casting all presuppositions about the world aside.
- **Hyperbole**: A hyperbole is an exaggeration. For example, "I'm so tired I could sleep for centuries."
- **Allusion**: An allusion is a reference to a character or event that happened in the past. T.S. Eliot's "The Waste Land" is a poem littered with allusions, including, "I will show you fear in a handful of dust," alluding to Genesis 3:19: "For you are dust, and to dust you shall return."
- **Pun**: Puns are used in popular culture to invoke humor by exploiting the meanings of words. In "Romeo and Juliet," Mercutio makes a pun after he is stabbed by Tybalt: "look for me tomorrow and you will find me a grave man."
- **Imagery**: This is a collection of images given to the reader by the author. If a text is rich in imagery, it is easier for the reader to imagine themselves in the author's world.

  One example of a poem that relies on imagery is William Carlos Williams' "The Red Wheelbarrow":

  > so much depends
  > upon
  >
  > a red wheel
  > barrow
  >
  > glazed with rain
  > water
  >
  > beside the white
  > chickens

  The starkness of the imagery and the placement of the words in this poem bring to life the images of a purely simple world. Through its imagery, this poem tells a story in sixteen words.

- **Symbolism**: A symbol is used to represent an idea or belief system. For example, poets in Western civilization have been using the symbol of a rose for hundreds of years to represent love. In Japan, poets have used the firefly to symbolize passionate love, and sometimes even spirits of those who have died. Symbols can also express powerful political commentary and can be used in propaganda.
- **Irony**: There are three types of irony: verbal, dramatic, and situational. Verbal irony is when a person states one thing and means the opposite. For example, a person is probably using irony when they say, "I can't wait to study for the exam next week." Dramatic irony occurs in a narrative and happens when the audience knows something that the characters do not. In the modern TV series *Hannibal*, the audience knows that Hannibal Lecter is a serial killer, but most of the main characters do not. Finally, situational irony is when one expects something to happen, and the opposite occurs. For example, we can say that a fire station burning down would be an instance of situational irony.

Idiomatic Expressions

Idiomatic expressions are phrases or groups of words that have an established meaning when used together that is unrelated to the literal meanings of the individual words. For example, consider the following sentence that includes a common idiomatic phrase:

> I know Phil is coming to visit this weekend because I heard it straight from the horse's mouth.

The speaker of this sentence did not consult a horse nor hear anything uttered from a horse in relation to Phil's visit. Instead, "straight from the horse's mouth" is an **idiom** that means the information came directly from an original or reliable source. As in the sentence above, it often means whatever said should be taken as truth because it was spoken by the person to which it pertains (in this case, Phil). The phrase is derived from the fact that sellers of horses at auctions would sometimes try to lie about the age of the horse. However, the size and shape of a horse's teeth can provide a fairly accurate estimate of the horse's true age. Therefore, the truth regarding the horse's age essentially comes straight from their mouth.

Finding errors in idiomatic expressions can be difficult because it requires familiarity with the idiom. Because there are more than one thousand idioms in the English language, memorizing all of them is impractical. However, it is recommended to review the most common ones. There are many webpages dedicated to listing and explaining frequently used idioms.

The test may ask you to identify errors in idiomatic expressions. These errors are typically one of two types. The idiomatic expression may be stated improperly, or it may be used in an incorrect context. In the first type of issue, the prepositions used are often incorrect. For example, it might say "straight *in* the horse's mouth" or "straight *with* the horse's mouth." In the second error type, the idiomatic expression is used improperly because the meaning it carries does not make sense in the context in which it appears.

Consider the following:

> He was looking straight from the horse's mouth when he complained about the phone his father bought him.

Here, the writer has confused the idiom "straight from the horse's mouth" with "looking a gift horse in the mouth," which means to find fault in a gift or favor.

Either type of error can be difficult to detect and correct without prior knowledge of the idiomatic phrase. Practicing usage of idioms and studying their origins can help you remember their meanings and precise wordings, which will then help you identify and correct errors in their usage.

Function of Transitional Words and Phrases

In writing, some sentences naturally lead to others, whereas in other cases, a new sentence expresses a new idea. Transitional phrases connect sentences and the ideas they convey, which makes the writing coherent. Transitional language also guides the reader from one thought to the next. For example, when pointing out an objection to the previous idea, starting a sentence with *however*, *but*, or *on the other hand* is transitional. When adding another idea or detail, writers use *also*, *in addition*, *furthermore*, *further*, *moreover*, *not only*, etc. Readers may have difficulty perceiving connections between ideas without such transitional wording.

Clerical

The clerical section of the exam tests your verbal, written, social, mathematical, and technical skills. It usually lasts for a few hours and may contain up to 100 questions.

Your administrative abilities will also be assessed. You may be asked to take a typing test to record your typing speed and accuracy. You may also be asked to properly file, code, or identify clerical errors. Putting information in the proper order is a large part of this section. You will be presented with data sets that appear similar, and you will need to conclude how they differ. Your ability to use a computer to perform basic data entry, respond to correspondence, schedule meetings, and provide customer support will be evaluated.

To determine your reading comprehension skills and verbal ability, you will be asked to read and answer questions about written passages. You may also be asked to define terms used in these passages. Your written communication abilities will be assessed, including your knowledge of English spelling, grammar, and syntax. You will be presented with a passage containing an error, and then you will have to select the correction from a list of answers. You may also be asked to reword sentences into coherent statements. Pay close attention to these passages. Even if you are an expert in the topic presented, your reading skills are being tested; so, read as if you are learning this information for the first time.

Your interpersonal skills will be measured as part of the social ability test. This is performed by asking questions that identify your personality type and your aptitude for rational decision making. The government is not looking for one particular personality, so relax and answer these questions honestly.

Many test takers worry about having their math skills scrutinized. However, take comfort in the fact that the hiring agency wants to ensure that you can solve basic math problems involving addition, subtraction, division, and multiplication. You'll also want to be familiar with fractions. You may be permitted to use a calculator during this part of the exam. Logic and reasoning will also be examined through word problems that require you to select the most practical answer from a list.

Your ability to think critically, communicate effectively, and perform basic administrative duties is crucial to becoming a civil servant. The exam may be on paper or computerized, and it will be administered by the governing agency to which you applied. You will be provided with ample time to complete each question. If you feel stuck, move on to the next question and come back later. Do not spend too much time fretting over one particular question.

A majority of the exam consists of multiple-choice questions. These tests vary, and there is no single, correct way to prepare for them. However, familiarizing yourself with the types of questions that may be included is one of the best things you can do.

Mathematics

Number Operations

Rational and Irrational Numbers

Rational numbers are those that can be written as a fraction, including **whole** or **negative** numbers, **fractions**, or **repeating decimals**. Examples of rational numbers include $\frac{1}{2}$, $\frac{5}{4}$, 2.75, and 8. Whole numbers can be written as fractions by putting the number itself as the numerator and 1 as the denominator; for example, 25 and 17 can be written as $\frac{25}{1}$ and $\frac{17}{1}$, respectively. One way of interpreting these fractions is to say that they are **ratios**, or comparisons of two quantities. The fractions given may represent 25 students to 1 classroom, or 17 desks to 1 computer lab. Repeating decimals can also be written as fractions; 0.3333 and 0.6666667 can be written as $\frac{1}{3}$ and $\frac{2}{3}$. Fractions can be described as having a part-to-whole relationship. The $\frac{1}{3}$ may represent 1 piece of pizza out of the whole cut into 3 pieces. The fraction $\frac{2}{3}$ may represent 2 pieces of the same whole pizza. Rational numbers, including fractions, can be added together. When the denominators of two fractions are the same, you can simply add the numerators. For example, adding the fractions $\frac{1}{3}$ and $\frac{2}{3}$ is as simple as adding the numerators, 1 and 2. The result is $\frac{3}{3}$, which equals 1. Both of these numbers are rational and represent a whole.

In addition to fractions, rational numbers also include whole numbers and negative integers. When whole numbers are added, the result is always greater than the addends (unless 0 is added to the number, in which case its value would remain the same). For example, the equation:

$$4 + 18 = 22$$

4 increased by 18 results in 22. When subtracting rational numbers, sometimes the result is a negative number. For example, the equation:

$$5 - 12 = -7$$

Taking 12 away from 5 results in a negative answer (− 7) because the starting number (5) is smaller than the number taken away (12). For multiplication and division, similar results are found. Multiplying rational numbers may look like the following equation:

$$5 \times 7 = 35$$

Both numbers are positive and whole, and the result is a larger number than the factors. The number 5 is counted 7 times, which results in a total of 35. Sometimes, the equation looks like:

$$-4 \times 3 = -12$$

The result is negative because a positive number times a negative number gives a negative answer. The rule is that any time a negative number and a positive number are multiplied or divided, the result is negative.

Operations with Rational Numbers

Basic Operations

The four basic operations include addition, subtraction, multiplication, and division. The result of addition is a **sum**, the result of subtraction is a **difference**, the result of multiplication is a **product**, and the result of division is a **quotient**. Each type of operation can be used when working with rational numbers.

These operations should first be learned using whole numbers. Addition needs to be done column by column. To add two whole numbers, add the ones column first, then the tens columns, then the hundreds, etc. If the sum of any column is greater than 9, a one must be carried over to the next column.

For example, the following is the result of $482 + 924$:

$$\begin{array}{r} \scriptstyle 1 \\ 482 \\ +924 \\ \hline 1406 \end{array}$$

Notice that the sum of the tens column was 10, so a one was carried over to the hundreds column. Subtraction is also performed column by column. Subtraction is performed in the ones column first, then the tens, etc. If the number on top is smaller than the number below, a one must be borrowed from the column to the left. For example, the following is the result of $5{,}424 - 756$:

$$\begin{array}{cccc} 4 & 13 & 11 & 14 \\ \not{5} & \not{4} & \not{2} & \not{4} \\ - & 7 & 5 & 6 \\ \hline 4 & 6 & 6 & 8 \end{array}$$

Notice that a one is borrowed from the tens, hundreds, and thousands place. After subtraction, the answer can be checked through addition. A check of this problem would be to show that:

$$756 + 4{,}668 = 5{,}424$$

In multiplication, the number on top is known as the **multiplicand,** and the number below is the **multiplier.** Complete the problem by multiplying the multiplicand by each digit of the multiplier. Make sure to place the ones value of each result under the multiplying digit in the multiplier. The final product is found by adding each partial product. The following example shows the process of multiplying 46 times 37:

$$\begin{array}{r} 46 \\ \times\ \ 7 \\ \hline 322 \end{array} \qquad \begin{array}{r} 46 \\ \times\ \ 3 \\ \hline 138 \end{array} \qquad \begin{array}{r} 46 \\ \times\ 37 \\ \hline 322 \\ 1380 \\ \hline 1702 \end{array}$$

Partial product

Partial product

Division can be performed using long division. When dividing, the first number is known as the **dividend,** and the second is the **divisor.** For example, with $a \div b = c$, a is the dividend, b is the divisor, and c is the quotient. For long division, place the dividend within the division bar and the divisor on the outside. For example, with $8{,}764 \div 4$, refer to the first problem in the diagram below. The first digit, 8, is divisible by 4 two times. Therefore, 2 goes above the division bar over the 8. Then, multiply 4 times 2 to get 8, and that product goes below the 8. Subtract to get 0, and then carry down the second digit, 7. 4 goes into 7 one time, with 3 left over. The 1 goes above the division bar over the 7. Then subtract the product of 4 times 1 (4) from 7 to get 3 and carry down the 6. Continuing this process for the next two digits results in a 9 and a 1. The final subtraction results in a 0, which means that 8,764 is evenly divisible by 4 with no remaining numbers.

The second example shows that:

$$4{,}536 \div 216 = 21$$

The steps are a little different because 216 cannot be contained in 4 or 5, so the first step is placing a 2 above the 3 because there are two 216's in 453.

Finally, the third example shows that:

$$546 \div 31 = 17 \text{ R19}$$

The 19 is a remainder. Notice that the final subtraction does not result in a 0, which means that 546 is not evenly divisible by 31. The remainder can also be written as a fraction over the divisor:

$$546 \div 31 = 17\frac{19}{31}$$

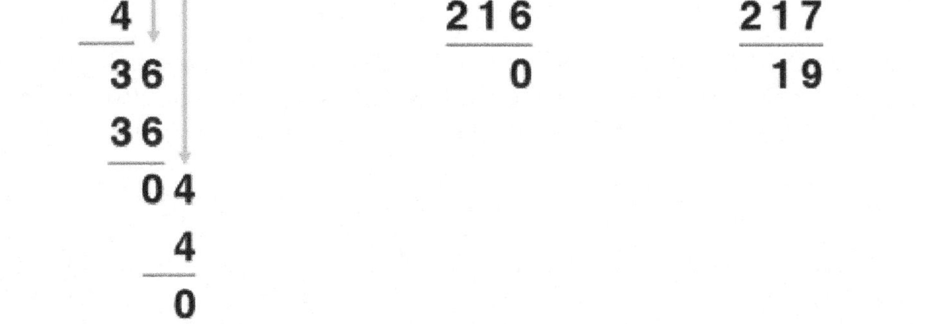

```
    2191            21          17 r 19
 4|8764      216|4536       31|546
   8             432            31
   07            216            236
    4            216            217
    36             0             19
    36
     04
      4
      0
```

A remainder can have meaning in a division problem with real-world application. Consider the third example above:

$$546 \div 31 = 17\,R19$$

Let's say that we had $546 to spend on calculators that cost $31 each, and we wanted to know how many we could buy. The division problem would answer this question. The result states that 17 calculators could be purchased with $19 left over. Note that the remainder will never be greater than or equal to the divisor.

Operations and Negative Numbers

Once the operations are understood with whole numbers, they can be used with negative numbers. There are many rules surrounding operations with negative numbers. First, consider addition. The sum of two numbers can be shown using a number line. For example, to add $-5 + (-6)$, plot the point -5 on the number line. Adding a negative number is the same as subtracting, so move 6 units to the left. This process results in landing on -11 on the number line, which is the sum of -5 and -6. If adding a positive number, move to the right on the number line. While visualizing this process using a number line is useful for understanding, it is more efficient to learn the rules of operations. When adding two numbers with the same sign, add the absolute values of both numbers, and use the common sign of both numbers for the sum. The absolute value simply means the positive form of the number. For example, to add $-5 + (-6)$, add their absolute values:

$$5 + 6 = 11$$

Then, introduce a negative symbol because both addends are negative. The result is -11. To add two integers with unlike signs, subtract the lesser absolute value from the greater absolute value, and apply the sign of the number with the greater absolute value to the result. For example, the sum $-7 + 4$ can be computed by finding the difference $7 - 4 = 3$ and then applying a negative because the value with the larger absolute value is negative. The result is -3. Similarly, the sum $-4 + 7$ can be found by computing the same difference but leaving it as a positive result because the addend with the larger absolute value is positive. Also, recall that any number plus 0 equals that number. This is known as the **Addition Property of 0**.

Subtracting two integers with opposite signs can be computed by changing to addition to avoid confusion. The rule is to add the first number to the opposite of the second number. The opposite of a number is the number with the same value on the other side of 0 on the number line. For example, -2 and 2 are opposites. Consider $4 - 8$. Change this to adding the opposite as follows: $4 + (-8)$. Then, follow the rules described above to obtain -4. Now consider $-8 - (-2)$. Change this problem to adding the opposite as $-8 + 2$, which equals -6. Notice that subtracting a negative number is really adding a positive number.

The operations of multiplication and division are actually less confusing than addition and subtraction because the rules are simpler to understand. If two factors in a multiplication problem have the same sign, the product is positive. If one factor is positive and one factor is negative, the product is negative. For example,

$$(-9)(-3) = 27 \text{ and } 9(-3) = -27$$

Also, a number multiplied by 0 always results in 0. If a problem consists of several multipliers, the result is negative if it contains an odd number of negative factors, and the result is positive if it contains an even number of negative factors. For example,

$$(-1)(-1)(-1)(-1) = 1 \text{ and } (-1)(-1)(-1)(-1)(-1) = -1$$

These two problems also display repeated multiplication, which can be written in a more compact notation using exponents. The first example can be written as $(-1)^4 = 1$, and the second example can be written as $(-1)^5 = -1$. Both are exponential expressions; -1 is the base in both instances, and 4 and 5 are the respective exponents. Note that a negative number raised to an odd power is always negative, and a negative number raised to an even power is always positive. Also, $(-1)^4$ is not the same as -1^4. In the first expression, the negative is included in the parentheses, but it is not in the second expression. The second expression is found by evaluating 1^4 first to get 1 and then applying the negative sign to obtain −1.

Similar rules apply within division. If two numbers in a division problem have the same sign, the quotient is positive. If two numbers in a division problem have different signs, the quotient is negative. For example:

$$14 \div (-2) = -7, and \quad -14 \div (-2) = 7$$

To check division, multiply the quotient by the divisor to obtain the dividend. Also, remember that 0 divided by any number is equal to 0. However, any number divided by 0 is undefined. It just does not make sense to divide a number by 0 parts.

Order of Operations

If more than one operation is to be completed in a problem, follow the **order of operations**. The mnemonic device, **PEMDAS**, states the order in which addition, subtraction, multiplication, and division need to be done. It also includes when to evaluate operations within grouping symbols and when to incorporate exponents. PEMDAS, which some remember by thinking "please excuse my dear Aunt Sally," refers to *parentheses*, *exponents*, *multiplication*, *division*, *addition*, and *subtraction*. First, complete any operation within parentheses or any other grouping symbol like brackets, braces, or absolute value symbols. Note that this does not refer to when parentheses are used to represent multiplication like $(2)(5)$. Then, any exponents must be computed. Next, multiplication and division are performed from left to right. Finally, addition and subtraction are performed from left to right. The following is an example in which the operations within the parentheses need to be performed first:

$$9-3(3^2-3+4\cdot 3)$$

$$\begin{aligned} &9-3(3^2-3+4\cdot 3) \quad \text{Work within the parentheses first} \\ &= 9-3(9-3+12) \\ &= 9-3(18) \\ &= 9-54 \\ &= -45 \end{aligned}$$

Operations and Decimals

Operations can be performed on rational numbers in decimal form. To write a fraction as an equivalent decimal expression, divide the numerator by the denominator. For example:

$$\frac{1}{8} = 1 \div 8 = 0.125$$

With the case of decimals, it is important to keep track of place value. To add decimals, make sure the decimal places are in alignment and add vertically. If the numbers do not line up because there are extra or missing place values in one of the numbers, then zeros may be used as placeholders. For example, $0.123 + 0.23$ becomes:

$$\begin{array}{r} 0.123 \\ +\underline{0.230} \\ 0.353 \end{array}$$

Subtraction is done the same way. Multiplication and division are more complicated. To multiply two decimals, place one on top of the other as in a regular multiplication process and do not worry about lining up the decimal points. Then, multiply as with whole numbers, ignoring the decimals. Finally, in the solution, insert the decimal point as many places to the left as there are total decimal values in the original problem. Here is an example of decimal multiplication:

$$\begin{array}{rl} 0.52 & \textit{2 decimal places} \\ \times\ \ 0.2 & \textit{1 decimal place} \\ \hline 0.104 & \textit{3 decimal places} \end{array}$$

The answer to 52 times 2 is 104, and because there are three decimal values in the problem, the decimal point is positioned three units to the left in the answer.

The decimal point plays an integral role throughout the whole problem when dividing with decimals. First, set up the problem in a long division format. If the divisor is not an integer, move the decimal to the right as many units as needed to make it an integer. The decimal in the dividend must be moved to the right the same number of places to maintain equality. Then, complete division normally. Below is an example of long division with decimals using the problem $12.72 \div 0.06$:

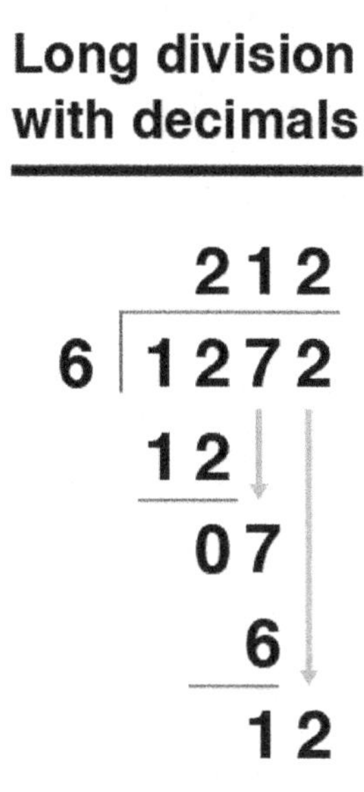

The decimal point in 0.06 needed to move two units to the right to turn it into an integer (6), so it also needed to move two units to the right in 12.72 to make it 1,272. The quotient is 212. To check a division problem, multiply the answer by the divisor to see if the result is equal to the dividend.

Operations and Fractions

When adding and subtracting fractions, the first step is to find the least common denominator. For example, the problem $\frac{3}{5}+\frac{6}{7}$ requires that a common multiple be found between 5 and 7. The smallest number that divides evenly by both 5 and 7 is 35. For the denominators to become 35, the first fraction must be multiplied by 7 and the second by 5. When $\frac{3}{5}$ is multiplied by 7, it becomes $\frac{21}{35}$. When $\frac{6}{7}$ is multiplied by 5, it becomes $\frac{30}{35}$. Once the fractions have the same denominator, the numerators can be added. The answer to the addition problem becomes:

$$\frac{3}{5}+\frac{6}{7}=\frac{21}{35}+\frac{30}{35}=\frac{41}{35}$$

The same technique can be used to subtract fractions. Multiplication and division may seem easier to perform because finding common denominators is unnecessary. If the problem reads $\frac{1}{3}\times\frac{4}{5}$, then the numerators and denominators are multiplied by each other and the answer is found to be $\frac{4}{15}$. For division, the problem must be changed to multiplication before performing operations. To complete a fraction division problem, you need to *leave*, *change*, and *flip* before multiplying. If the problem reads $\frac{3}{7}\div\frac{3}{4}$, then the first fraction is *left* alone, the operation is *changed* to multiplication, and the last fraction is *flipped*. The problem becomes:

$$\frac{3}{7}\times\frac{4}{3}=\frac{12}{21}$$

Root Operations

Another operation that can be performed on rational numbers is the **square root**. Dealing with real numbers only, the **positive square root** of a number is the factor that, when multiplied by itself, results in the original number. For example,

$$\sqrt{49}=\sqrt{7\times 7}=7$$

A **perfect square** is a number that has a whole number as its square root. Examples of perfect squares are 1, 4, 9, 16, 25, etc. If a number is not a perfect square, an approximation can be found with a calculator. For example, $\sqrt{67}=8.185$, rounded to the nearest thousandth place. Taking the square root of a fraction that includes perfect squares involves breaking up the problem into the square root of the numerator separate from the square root of the denominator.

For example:

$$\sqrt{\frac{16}{25}}=\frac{\sqrt{16}}{\sqrt{25}}=\frac{4}{5}$$

If the fraction does not contain perfect squares, a calculator can be used. For example, $\sqrt{\frac{2}{5}}=0.632$, rounded to the nearest thousandth place. A common application of square roots involves the **Pythagorean theorem**. In a right triangle, the sum of the squares of the two legs equals the square of the hypotenuse.

For example, consider the following right triangle:

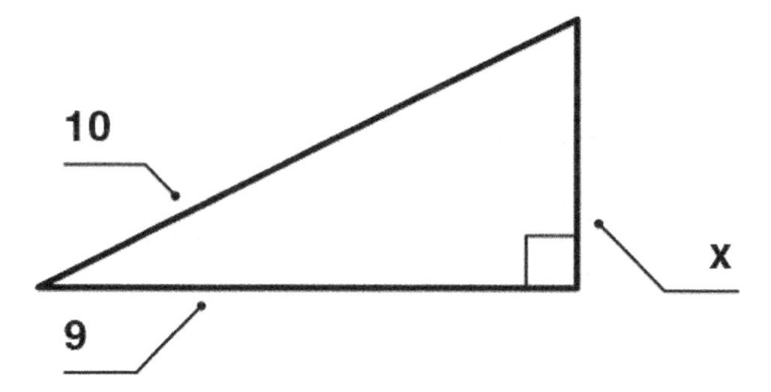

The missing side, *x*, can be found using the Pythagorean theorem.

$$9^2 + x^2 = 10^2$$

$$81 + x^2 = 100$$

$$x^2 = 19$$

To solve for *x*, take the square root of both sides. Therefore, $x = \sqrt{19} = 4.36$, which has been rounded to two decimal places.

In addition to the square root, the cube root is another operation. If a number is a **perfect cube**, the cube root of that number is a factor that, when cubed, results in the original number. For example:

$$\sqrt[3]{27} = \sqrt[3]{3 \times 3 \times 3} = 3$$

A negative number has a *cube root*, which will also be a negative number. For example:

$$\sqrt[3]{-27} = \sqrt[3]{(-3)(-3)(-3)} = -3$$

Similar to square roots, if the number is not a perfect cube, a calculator can be used to find an approximation. Therefore, $\sqrt[3]{\frac{2}{3}} = 0.873$, rounded to the nearest thousandth place.

The number relating to the root is known as the **index.** Given the following root, $\sqrt[3]{64}$, 3 is the index, and 64 is the **radicand.** The entire expression is known as the **radical.** Higher-order roots exist when the index is larger than 3. They can be broken up into two groups: even and odd roots. **Even roots**, when the index is an even number, follow the properties of square roots. They are found by finding the number that, when multiplied by itself the number of times indicated by the index, results in the radicand. Negative numbers cannot have an even root. For example, the 4th root of 81 is equal to 3 because $3^4 = 81$. This radical is written as:

$$\sqrt[4]{81} = 3$$

Odd roots, when the index is an odd number, follow the properties of cube roots. A negative number has an odd root. Similarly, an odd root is found by finding the number that, when multiplied by itself the number of times indicated by the index, gives the radicand.

For example, the fifth root of 32 is equal to 2 because:

$$\sqrt[5]{32} = \sqrt[5]{2 \times 2 \times 2 \times 2 \times 2} = 2$$

Rewriting Expressions Involving Radicals and Rational Exponents

Rational exponents represent one way to show how roots are used to express multiplication of any number by itself. For example, $3^{\frac{2}{3}}$ has a base of 3 and rational exponent of $\frac{2}{3}$. It can be rewritten as the cube root of 3 raised to the second power, or $\sqrt[3]{3^2}$. Any number with a rational exponent can be written this way. The numerator becomes the root, and the denominator becomes the whole number exponent. Another example is $4^{\frac{3}{2}}$. It can be rewritten as the square root of four to the third power, or $\sqrt[2]{4^3}$. To simplify this, first solve for 4 to the third power:

$$4^3 = 4 \times 4 \times 4 = 64$$

Then take the square root of 64 ($\sqrt[2]{64}$), which yields an answer of 8.

The *n*th root of *a* is given as $\sqrt[n]{a}$, which is called a **radical.** Typical values for *n* are 2 and 3, which represent the square and cube roots. In this form, *n* represents an integer greater than or equal to 2, and a is a real number. If *n* is even, *a* must be nonnegative, and if *n* is odd, *a* can be any real number. This radical can be written in exponential form as $a^{\frac{1}{n}}$. Therefore, $\sqrt[4]{15}$ is the same as $15^{\frac{1}{4}}$ and $\sqrt[3]{-5}$ is the same as $(-5)^{\frac{1}{3}}$.

In a similar fashion, the *n*th root of *a* can be raised to a power *m*, which is written as $\left(\sqrt[n]{a}\right)^m$. This expression is the same as $\sqrt[n]{a^m}$. For example:

$$\sqrt[2]{4^3} = \sqrt[2]{64} = 8 = \left(\sqrt[2]{4}\right)^3 = 2^3$$

Because $\sqrt[n]{a} = a^{\frac{1}{n}}$, both sides can be raised to an exponent of *m*, resulting in:

$$\left(\sqrt[n]{a}\right)^m = \sqrt[n]{a^m} = a^{\frac{m}{n}}$$

This rule allows:

$$\sqrt[2]{4^3} = \left(\sqrt[2]{4}\right)^3 = 4^{\frac{3}{2}} = (2^2)^{\frac{3}{2}} = 2^{\frac{6}{2}} = 2^3 = 8$$

Negative exponents can also be incorporated into these rules. Any time an exponent is negative, the base expression must be flipped to the other side of the fraction bar and rewritten with a positive exponent. For instance,

$$2^{-3} = \frac{1}{2^3} = \frac{1}{8}$$

Therefore, two more relationships between radical and exponential expressions are:

$$a^{-\frac{1}{n}} = \frac{1}{\sqrt[n]{a}} \text{ and } a^{-\frac{m}{n}} = \frac{1}{\sqrt[n]{a^m}} = \frac{1}{\left(\sqrt[n]{a}\right)^m}$$

Thus:

$$8^{-3} = \frac{1}{\sqrt[3]{8}} = \frac{1}{2}$$

All of these relationships are very useful when simplifying complicated radical and exponential expressions. If an expression contains both forms, use one of these rules to change the expression to contain either all radicals or all exponential expressions. This process makes the entire expression much easier to work with, especially if the expressions are contained within equations.

Consider the following example: $\sqrt{x} \times \sqrt[4]{x}$. It is written in radical form; however, it can be simplified into one radical by using exponential expressions first. The expression can be written as $x^{\frac{1}{2}} \times x^{\frac{1}{4}}$. It can be combined into one base by adding the exponents as:

$$x^{\frac{1}{2}+\frac{1}{4}} = x^{\frac{3}{4}}$$

Writing this back in radical form, the result is $\sqrt[4]{x^3}$.

Solving Problems Using Scientific Notation

Scientific notation is a system used to represent numbers that are very large or very small. Sometimes numbers are way too big or small to be written out with all their digits, so scientific notation is used as a way to express these numbers in a simpler way.

Scientific notation takes a number's decimal notation and turns it into exponent form, as shown in the table below:

Decimal Notation	Scientific Notation
5	5×10^0
500	5×10^2
$10{,}000{,}000$	1×10^7
$8{,}000{,}000{,}000$	8×10^9
$-55{,}000$	-5.5×10^4
$.00001$	10^{-5}

Let's say you have the number 125,000. You would write this in scientific notation as 1.25×10^5. Multiplying by 10^5 essentially means moving the decimal point to the right five times. Therefore, writing in scientific notation allows you to write the number in three decimal places instead of five. Another example, as shown in the table above, is the number .00001. In scientific notation, it becomes 10^{-5}. The negative exponent means that you move the decimal point to the left five times.

Rounding

Sometimes it is helpful to round answers that are in decimal form. First, find the place value to which the rounding needs to be done. Then, look at the digit to the right of it. If that digit is 4 or less, the number in the place value to its left stays the same, and everything to its right becomes a 0. This process is known as **rounding down.** If that digit is 5 or higher, the number in the place value to its left increases by 1, and every number to its right becomes a 0. This is called **rounding up**. Excess 0s at the end of a decimal can be dropped. For example, 0.145 rounded to the nearest hundredth place would be rounded up to 0.15, and 0.145 rounded to the nearest tenth place would be rounded down to 0.1.

When performing operations with rational numbers, it might be helpful to round the numbers in the original problem to get a rough idea of what the answer should be. **Front-end rounding** might also be helpful in many situations. In this type of rounding, the first digit of a number is rounded up to the highest possible place value. Then, all digits following the first become 0. For example, 267 would be rounded to 300, and 5,321 would be rounded to 6,000.

Applying Estimation Strategies and Rounding Rules to Real-World Problems

Sometimes it is helpful to find an estimated answer to a problem rather than working out an exact answer. An **estimation** might be much quicker to find, and it might be all that is required given the scenario. For example, if Aria goes grocery shopping with only $100, she might want to estimate the total of the items she is purchasing to determine if she has enough money to cover them. Also, an estimation can help determine if an answer makes sense. For instance, if you estimate that an answer should be in the 100s, but your result is a fraction less than 1, something is probably wrong in the calculation.

The first type of estimation involves rounding. As mentioned, **rounding** consists of expressing a number in terms of the nearest place value. For instance, 1,654.2674 rounded to the nearest thousand is 2,000, and the same number rounded to the nearest thousandth is 1,654.267. Rounding can make it easier to estimate totals at the store. For example, a can of corn that costs $0.79 can be rounded to $1.00, and then all other items can be rounded in a similar manner and added together.

When working with larger numbers, it might make more sense to round to higher place values. For example, when estimating the total value of a car dealership's inventory, it would make sense to round the car values to the nearest thousands place. The price of a car that is on sale for $15,654 can be estimated at $16,000. All other cars on the lot could be rounded in the same manner and then added together. Depending on the situation, it might make sense to calculate an over-estimate. For example, to make sure Aria has enough money at the grocery store, rounding up for each item would ensure that she has enough money when it comes time to pay. A $0.40 item rounded up to $1.00 would ensure that there is a dollar to cover that item. Traditional rounding rules would round $0.40 to $0, which does not make sense in this particular real-world setting. Aria might not have a dollar available at checkout to pay for that item if she uses traditional rounding. It is up to the customer to decide the best approach when estimating.

Estimating is also very helpful when working with measurements. Bryan is updating his kitchen and wants to retile the floor. Again, an over-measurement might be useful. For instance, one side of the kitchen might have an exact measurement of 14.32 feet, and the most useful measurement needed to buy tile could be estimating this quantity to be 14.5 feet. If the kitchen was rectangular and the other side measured 10.9 feet, Bryan might round the other side to 11 feet. Therefore, Bryan would find the total tile necessary according to the following area calculation:

$$14.5 \times 11 = 159.5 \text{ square feet}$$

To make sure he purchases enough tile, Bryan would probably want to purchase at least 160 square feet of tile. This is a scenario in which an estimation might be more useful than an exact calculation. Having more tile than necessary is better than having an exact amount, in case any tiles are broken or otherwise unusable.

Estimates can also be helpful with square roots. If the square root of a number is unknown, then you can use the closest perfect square to help you approximate. For example, $\sqrt{50}$ is not equal to a whole number, but 50 is close to 49, which is a perfect square. Since $\sqrt{49} = 7$, that means $\sqrt{50}$ must be a little bit larger than 7. The actual approximation, rounded to the nearest thousandth, is 7.071.

Finally, estimation is helpful when exact answers are necessary. Consider a situation in which Sabina has many operations to perform on numbers with decimals, and she is allowed a calculator to find the result. Even though an exact result can be obtained with a calculator, there is always a possibility that Sabina could make an error while inputting the data. For example, she could miss a decimal place, or misuse parentheses, causing a problem with the actual order of operations. A quick estimation at the beginning could help ensure that her final answer is within the correct range. Sabina has to find the exact total of 10 cars listed for sale at the dealership. Each price has two decimal places included to account for both dollars and cents. If one car is listed at $21,234.43 but Sabina incorrectly inputs into the calculator the price of $2,123.443, this error would throw off the final sum by almost $20,000. If her final calculation was significantly different from her estimation, then she would know to look for an error.

Reasoning Quantitatively and Using Units to Solve Problems

Converting Within and Between Standard and Metric Systems

When working with dimensions, sometimes the given units do not match the formula, and a conversion is required. **The metric system's** base units are *meter* for length, *gram* for mass, and *liter* for liquid volume. This system expands to three places above the base unit and three places below. These places correspond to prefixes that each signify a specific base of 10.

The following table shows the conversions:

kilo-	hecto-	deka-	base	deci-	centi-	milli-
1,000 times the base	100 times the base	10 times the base		1/10 times the base	1/100 times the base	1/1000 times the base

To convert between units within the metric system, values with a base ten can be multiplied. For example, 3 meters is equivalent to .003 kilometers. The decimal moved three places (the same number of zeros for kilo-) to the left (the same direction from base to kilo-). Three meters is also equivalent to 3,000 millimeters. The decimal is moved three places to the right because the prefix milli- is three places to the right of the base unit.

The English Standard system, which is used in the United States, uses the base units of *foot* for length, *pound* for weight, and *gallon* for liquid volume. Conversions within the English Standard system are not as easy as those within the metric system because the former does not use a base ten model. The following table shows the conversions within this system.

Length	Weight	Capacity
1 foot (ft) = 12 inches (in) 1 yard (yd) = 3 feet 1 mile (mi) = 5280 feet 1 mile = 1760 yards	1 pound (lb) = 16 ounces (oz) 1 ton = 2000 pounds	1 tablespoon (tbsp) = 3 teaspoons (tsp) 1 cup (c) = 16 tablespoons 1 cup = 8 fluid ounces (oz) 1 pint (pt) = 2 cups 1 quart (qt) = 2 pints 1 gallon (gal) = 4 quarts

When converting within the English Standard system, most calculations include a conversion to the base unit and then another to the desired unit. For example, take the following problem: $3\ quarts = \_\_\_\ cups$. There is no straight conversion from quarts to cups, so the first conversion is from quarts to pints. There are 2 pints in 1 quart, so there are 6 pints in 3 quarts.

This conversion can be solved as a proportion:

$$\frac{3\ qt}{x} = \frac{1\ qt}{2\ pints}$$

It can also be observed as a ratio, 2:1, expanded to 6:3. Then the 6 pints must be converted to cups. The ratio of pints to cups is 1:2, so the expanded ratio is 6:12. Therefore, 6 pints equals 12 cups.

Another way to solve this problem would be to set up a series of fractions and cancel out units. Consider the following expression:

$$\frac{3\ quarts}{1} \times \frac{2\ pints}{1\ quart} \times \frac{2\ cups}{1\ pint}$$

It is set up so that units on the top and bottom cancel each other out.

$$\frac{3\ \cancel{quarts}}{1} \times \frac{2\ \cancel{pints}}{1\ \cancel{quart}} \times \frac{2\ cups}{1\ \cancel{pint}}$$

The numbers can be calculated as $3 \times 3 \times 2$ on the top and 1 on the bottom. It still yields an answer of 12 cups.

This process of setting up fractions and canceling out matching units can be used for conversions between standard and metric systems. A few common equivalent conversions are 2.54 cm = 1 inch, 3.28 feet = 1 meter, and 2.205 pounds = 1 kilogram. Writing these as fractions allows them to be used in conversions. For the problem 5 meters = ___ feet, use the feet-to-meter conversion and start with the expression $\frac{5\ meters}{1} \times \frac{3.28\ feet}{1\ meter}$. The "meters" will cancel each other out, leaving "feet" as the final unit. Calculating the numbers yields 16.4 feet. This problem only required two fractions. Others may require longer expressions, but the underlying rule stays the same. When a unit in the numerator of a fraction matches a unit in the denominator, then they cancel each other out. Using this logic and the conversions given above, many units can be converted between and within the different systems.

The conversion between Fahrenheit and Celsius is found in a formula:

$$°C = (°F - 32) \times \frac{5}{9}$$

For example, to convert 78°F to Celsius, the given temperature would be entered into the formula:

$$°C = (78 - 32) \times \frac{5}{9}$$

Solving the equation, the temperature comes out to be 25.56°C. To convert in the other direction, the formula becomes:

$$°F = °C \times \frac{9}{5} + 32$$

Remember the order of operations when calculating these conversions.

Choosing a Level of Accuracy Appropriate to Limitations on Measurement

Accuracy is defined as the closeness of a given measurement to the actual value of a certain object. The accuracy of a measurement depends on both the tools used to obtain the measurement and the units that are provided. For example, if you have a scale at home that does not utilize any decimal values, it actually rounds your weight to the nearest whole number. Therefore, it might say that you weigh 125 pounds, when your true weight is 125.4 pounds. The desired accuracy depends on the situation. For instance, you might not care that your weight is rounded to the nearest whole number. However, if you are using that scale to weigh your luggage, and it states that your bag weighs 50 pounds when it actually weighs 50.4 pounds, then you might end up paying an overweight fee at the airport. The difference between the two weights has to do with precision. **Precision** refers to the exactness of measurements, or how close they are to one another. For example, if you step on the scale numerous times and get weights of 125, 127, and 123 pounds, the scale is not that precise. It is important to note that measurements can be precise, but not accurate. Imagine using a scale that is improperly calibrated. You might weigh yourself five times and get 140 pounds every time. The measurements may be precise, but if your true weight is 125 pounds, then they were not very accurate.

A way to discuss accuracy in more detail is to use both absolute and relative accuracy. **Absolute accuracy** is the absolute value of the difference between an exact measurement and the approximate measurement. For instance, in the examples above, the absolute accuracy was 0.4 because the actual and measured amounts differed by 0.4 pounds. **Relative accuracy** equals the absolute accuracy divided by the exact value, turned into a percent. Therefore, the relative accuracy of your measurement was:

$$\frac{0.4}{125.4} = 0.32\%$$

The relative accuracy of your cat's measurement was:

$$\frac{0.4}{50.4} = 0.79\%$$

Solving Multi-Step Problems Involving Rational Numbers and Proportional Relationships

Solving Problems Involving Ratios and Rates of Change

A **ratio** is the comparison of two different quantities. Comparing 2 apples to 3 oranges results in the ratio 2:3, which can be expressed as the fraction $\frac{2}{3}$. Note that order is important when discussing ratios. The number mentioned first becomes the **numerator**, and the number mentioned second becomes the **denominator**. The ratio 2:3 does not mean the same thing as 3:2. Also, if the ratios have units attached to them, the two quantities must use the same units. For example, to compare 8 feet to 4 yards, it would make sense to convert 4 yards to feet by multiplying by 3. Therefore, the ratio would be 8 feet to 12 feet, which can be expressed as the fraction $\frac{8}{12}$. Note that it is proper to refer to ratios in lowest terms. Therefore, the ratio of 8 feet to 4 yards is equivalent to the fraction $\frac{2}{3}$.

Many real-world problems involve ratios. Consider an amusement park that sold 345 tickets last Saturday. If 145 tickets were sold to adults and the rest of the tickets were sold to children, what would the ratio of the number of adult tickets to children's tickets be? A common mistake would be to say the ratio is 145:345. However, 345 is the total number of tickets sold. There were $345 - 145 = 200$ tickets sold to children. The correct ratio of adult to children's tickets is 145:200. As a fraction, this expression is written as $\frac{145}{200}$, which can be reduced to $\frac{29}{40}$.

While a ratio compares two measurements using the same units, **rates** compare two measurements with different units. Examples of rates would be $200 for 8 hours of work, or 500 miles per 20 gallons of gas. Because the units are different, it is important to always include the units when discussing rates. Key words in rate problems include *for, per, on, from,* and *in.* Just as with ratios, it is important to write rates in lowest terms. A common rate in real-life situations is cost per unit, which describes how much one item/unit costs. When evaluating the cost of an item that comes in several sizes, the cost per unit rate can help buyers determine the best deal. For example, if 2 quarts of soup was sold for $3.50 and 3 quarts was sold for $4.60, to determine the best buy, the cost per quart should be found.

$$\frac{\$3.50}{2} = \$1.75\, per\ quart$$

and

$$\frac{\$4.60}{3} = \$1.53\, per\ quart$$

Therefore, the better deal would be the 3-quart option.

Rate of change problems involve calculating a quantity per some unit of measurement. Usually, the unit of measurement is time. For example, meters per second is a common rate of change. To calculate this measurement, find the amount traveled in meters and divide by total time traveled. The result is the average speed over the entire time interval. Another common rate of change is miles per hour. Consider the following problem that involves calculating an average rate of change in temperature. Last Saturday, the temperature at 1:00 a.m. was 34 degrees Fahrenheit, and at noon, the temperature had increased to 75 degrees Fahrenheit. What was the average rate of change over that time interval? The average rate of change is calculated by finding the change in temperature and dividing that number by the total hours elapsed.

Therefore, the rate of change was equal to:

$$\frac{75 - 34}{12 - 1} = \frac{41}{11} degrees\ per\ hour$$

This quantity rounded to two decimal places is equal to 3.73 degrees per hour.

A common rate of change that appears in algebra is the slope calculation. In the equation for a straight line, $y = mx + b$, the slope, *m,* is equal to:

$$\frac{rise}{run} \text{ or } \frac{change\ in\ y}{change\ in\ x}$$

In other words, **slope** is equivalent to the ratio of the vertical and horizontal changes between any two points on a line. The vertical change is known as the **rise**, and the horizontal change is known as the **run**. Given any two points on a line (x_1, y_1) and (x_2, y_2), slope can be calculated with the formula:

$$m = \frac{y_2 - y_1}{x_2 - x_1} = \frac{\Delta y}{\Delta x}$$

Common real-world applications of slope include determining how steep a staircase should be, calculating how steep a road is, and determining how to build a wheelchair ramp.

Many times, problems involving rates and ratios involve proportions. A **proportion** states that two ratios (or rates) are equal. Cross multiplication can be used to determine if a proportion is true, meaning both ratios are equivalent. To cross multiply, first multiply the top of the first fraction and the bottom of the second, then multiply the top of the second fraction and the bottom of the first. For example, $\frac{4}{40} = \frac{1}{10}$ grants the cross product $4 \times 10 = 40 \times 1$. This results in $40 = 40$, showing that the proportion is true. Cross products are used when proportions are involved in real-world problems. Consider the following: if 3 pounds of fertilizer will cover 75 square feet of grass, how many pounds are needed for 375 square feet? To solve this problem, set up a proportion using two ratios. Let x equal the unknown quantity, pounds needed for 375 feet. Setting the two ratios equal to one another yields the equation $\frac{3}{75} = \frac{x}{375}$. Cross multiplication gives:

$$3 \times 375 = 75x$$

Therefore, $1{,}125 = 75x$. Divide both sides by 75 to get $x = 15$. Therefore, 15 pounds of fertilizer is needed to cover 375 square feet of grass.

Another application of proportions involves similar triangles. If two triangles have corresponding angles with the same measurements and corresponding sides with proportional measurements, the triangles are said to be **similar**. If two are the same, the third pair of angles are equal as well because the sum of all angles in a triangle is equal to 180 degrees.

For example, consider the following set of similar triangles:

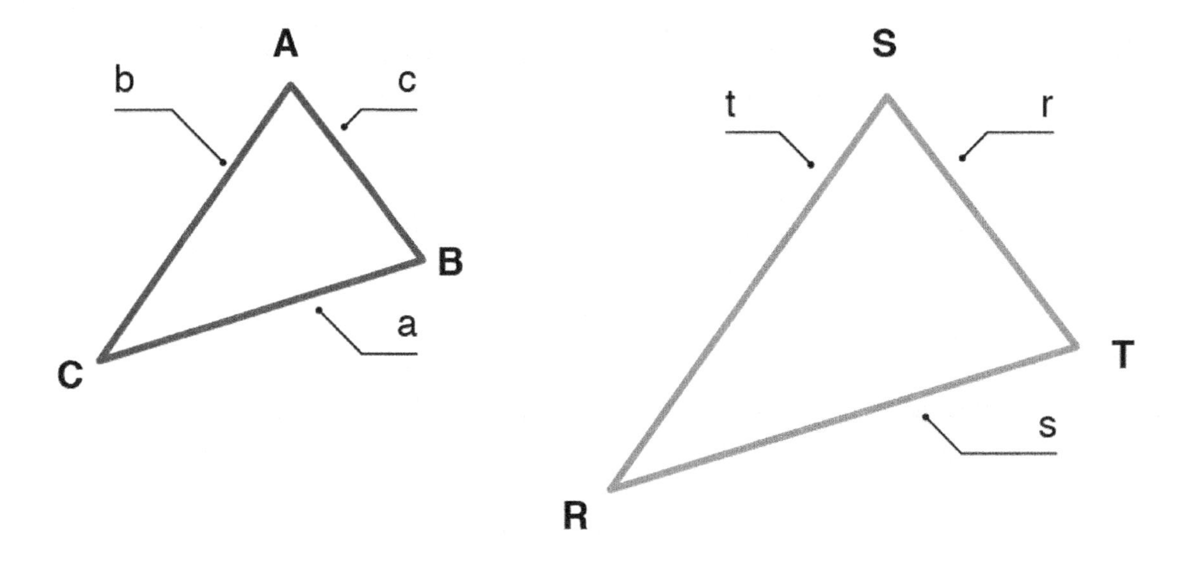

Angles A and S have the same measurement, angles C and R have the same measurement, and angles B and T have the same measurement.

Therefore, the following proportion can be set up from the sides:

$$\frac{a}{s} = \frac{c}{r} = \frac{b}{t}$$

This proportion can be helpful in finding missing lengths in pairs of similar triangles. For example, if the following triangles are similar, a proportion can be used to find the missing side lengths, *a* and *b*.

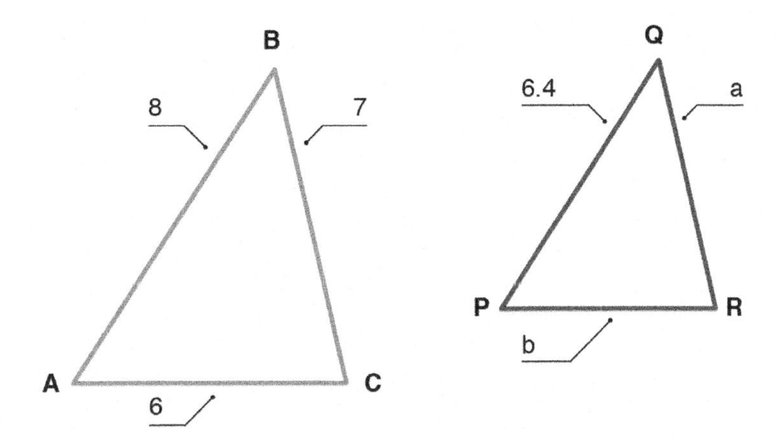

The proportions $\frac{8}{6.4} = \frac{6}{b}$ and $\frac{8}{6.4} = \frac{7}{a}$ can both be cross multiplied and solved to obtain *a* = 5.6 and *b* = 4.8.

A real-life situation that uses similar triangles involves measuring shadows to find heights of unknown objects. Consider the following problem: A building casts a shadow that is 120 feet long, and at the same time, another building that is 80 feet high casts a shadow that is 60 feet long. How tall is the first building? Each building, together with the sun rays and shadows casted on the ground, forms a triangle. They are similar because each building forms a right angle with the ground, and the sun rays form equivalent angles. Therefore, these two pairs of angles are both equal. Because all angles in a triangle add up to 180 degrees, the third angles are equal as well. Both shadows form corresponding sides of the triangle, the buildings form corresponding sides, and the sun rays form corresponding sides. Therefore, the triangles are similar, and the following proportion can be used to find the missing building length:

$$\frac{120}{x} = \frac{60}{80}$$

Cross multiply to obtain the equation $9600 = 60x$. Then, divide both sides by 60 to obtain $x = 160$. This means that the first building is 160 feet high.

Solving Problems with Percentages

Percentages are defined as parts per one hundred. To convert a decimal to a percentage, move the decimal point two units to the right and place the percent sign after the number. Percentages appear in many scenarios in the real world. It is important to make sure the statement containing the percentage is translated to a correct mathematical expression. Be aware that it is extremely common to make a mistake when working with percentages within word problems.

The following is an example of a word problem containing a percentage: 35% of people speed when driving to work. In a group of 5,600 commuters, how many would be expected to speed on the way to their place of employment? The answer to this problem is found by finding 35% of 5,600. First, change the percentage to the decimal 0.35. The word "of" indicates multiplication, so the equation would be $0.35 \times 5{,}600 = 1{,}960$. Therefore, it would be expected that 1,960 of those commuters would speed on their way to work based on the data given.

The following problem illustrates another way to use percentages: *Teachers work 8 months out of the year. What percent of the year do they work?* To answer this problem, find what percent *of* 12 the number 8 *is*, because there are 12 months in a year. To do this, divide 8 by 12, and convert that number to a percentage:

$$\frac{8}{12} = \frac{2}{3} = 0.66\bar{6}$$

The percentage rounded to the nearest tenth place tells us that teachers work 66.7% of the year.

Percentage problems can also find missing quantities like in the following question: 60% of what number is 75? To find the missing quantity, turn the question into an equation. Let x be equal to the missing quantity. Therefore, $0.60x = 75$. Divide each side by 0.60 to obtain 125. Therefore, 60% of 125 is equal to 75.

Sales tax is an important application of percentages because tax rates are usually given as percentages. For example, a city might have an 8% sales tax rate. Therefore, when an item is purchased with that tax rate, the real cost to the customer is 1.08 times the price in the store. For example, a $25 pair of jeans costs the customer $\$25 \times 1.08 = \$27$. If the sales tax rate is unknown, it can be determined after an item is purchased. If a customer visits a store and purchases an item for $21.44, but the price in the store was $19, they can find the tax rate by first subtracting $21.44 – $19 to obtain $2.44, the sales tax amount. The sales tax is a percentage of the in-store price. Therefore, the tax rate is $\frac{2.44}{19} = 0.128$, which has been rounded to the nearest thousandths place. In this scenario, the actual sales tax rate given as a percentage is 12.8%.

Algebra

Interpreting Parts of an Expression

An **algebraic expression** is a mathematical phrase that may contain numbers, variables, and mathematical operations. An expression represents a single quantity. For example, $3x + 2$ is an algebraic expression.

An **algebraic equation** is a mathematical sentence with two expressions that are equal to each other. That is, an equation must contain an *equals* sign, as in $3x + 2 = 17$. This statement says that the value of the expression on the left side of the equation is equivalent to the value of the expression on the right side. The equals sign (=) is the difference between an expression and an equation.

Example: Determine whether each of these is an expression or an equation.

a. $16 + 4x = 9x - 7$ Solution: Equation

b. $-27x - 42 + 19y$ Solution: Expression

c. $4 = x + 3$ Solution: Equation

In an algebraic expression, the **variables,** such as x and y, are the unknowns. To add and subtract linear algebraic expressions, you must combine like terms. **Like terms** are terms that have the same variable with the same exponent. Terms without a variable component are called **constants**.

Performing Arithmetic Operations on Polynomials and Rational Expressions

Simplifying Expressions

A **rational expression** is a fraction in which both the numerator and denominator are polynomials that are not equal to zero. A **polynomial** is a mathematical expression containing addition, subtraction, or multiplication of one or more constants multiplied by variables raised to positive powers. Here are some examples of rational expressions:

$$\frac{2x^2 + 6x}{x}$$

$$\frac{x - 2}{x^2 - 6x + 8}$$

$$\frac{x^3 - 1}{x + 2}$$

Such expressions can be simplified using different forms of division. The first example can be simplified in two ways. First, because the denominator is a monomial (one term), the expression can be split up into two expressions:

$$\frac{2x^2}{x} + \frac{6x}{x}$$

Then, cancelling out an x in each numerator and the x in each denominator results in $2x + 6$. Second, it can be simplified by factoring and then crossing common factors out of the numerator and denominator. For instance:

$$\frac{2x(x + 3)}{x} = 2(x + 3)$$

$$2x + 6$$

The second expression above can also be simplified using factoring. It can be written as:

$$\frac{x - 2}{(x - 2)(x - 4)} = \frac{1}{x - 4}$$

Finally, the third example can only be simplified using long division, as there are no common factors in the numerator and denominator. First, divide the first term of the numerator by the first term of the denominator, then write the result in the quotient. Then, multiply the divisor by that number and write it below the dividend. Subtract and continue the process until each term in the divisor is accounted for. Here is the actual long division:

Simplifying Expressions Using Long Division

$$
\begin{array}{r|cccc}
 & x^2 & -2x & +4 & \\
\hline
x+2 & x^3 & & & -1 \\
 & x^3 & +2x^2 & & \\
\hline
 & & -2x^2 & & -1 \\
 & & -2x^2 & -4x & \\
\hline
 & & & 4x & -1 \\
 & & & 4x & +8 \\
\hline
 & & & & -9
\end{array}
$$

Adding and Subtracting Expressions

As mentioned previously, you must combine like terms to add and subtract linear algebraic expressions. In the following example, the x-terms can be added because the variable and exponent are the same. The other like terms are **constants** because they have no variable component.

Example: Add $(3x - 5) + (6x + 14)$

$3x - 5 + 6x + 14$	*Rewrite without parentheses*
$3x + 6x - 5 + 14$	*Commutative Property of Addition*
$9x + 9$	*Combine like terms*

When subtracting linear expressions, be careful to add the opposite when combining like terms. Do this by distributing -1, which means multiplying each term inside the second parenthesis by -1. Remember that doing this changes the sign of each term.

Example: Subtract $(17x + 3) - (27x - 8)$

$17x + 3 - 27x + 8$	*Distributive Property*
$17x - 27x + 3 + 8$	*Commutative Property of Addition*
$-10x + 11$	*Combine like terms*

Example: Simplify by adding or subtracting:

$(6m + 28z - 9) + (14m + 13) - (-4z + 8m + 12)$

$6m + 28z - 9 + 14m + 13 + 4z - 8m - 12$ *Distributive Property*

$6m + 14m - 8m + 28z + 4z - 9 + 13 - 12$ *Commutative Property of Addition*

$12m + 32z - 8$ *Combine like terms*

Writing Expressions in Equivalent Forms to Solve Problems

Two algebraic expressions are equivalent if they represent the same value, even if they look different. To obtain an equivalent form of an algebraic expression, follow the laws of algebra. For instance, addition and multiplication are both commutative, meaning these operations can be performed in any order. For instance, $4x + 2y$ is equivalent to $2y + 4x$ and $(y \times 2) + (x \times 4)$. Also, the distributive law allows a number to be distributed throughout parentheses, as in the following:

$$a(b + c) = ab + ac$$

The expressions on both sides of the equals sign are equivalent. Collecting like terms is also important when working with equivalent forms because the simplest version of an expression is always the easiest one to work with.

An expression is not an equation; therefore, expressions cannot undergo multiplication, division, addition, or subtraction and still have equivalent expressions. These processes can only happen in equations when the same step is performed on both sides of the equals sign.

Using the Distributive Property to Generate Equivalent Linear Algebraic Expressions

The Distributive Property:

$$a(b + c) = ab + ac$$

The **distributive property** is a way of taking a factor and multiplying it through a given expression in parentheses. Each term inside the parentheses is multiplied by the outside factor, eliminating the parentheses.

Example: Use the distributive property to write an equivalent algebraic expression:

$3(2x + 7y + 6)$

$3(2x) + 3(7y) + 3(6)$ *Distributive property*

$6x + 21y + 18$ *Simplify*

Because $a - b$ can be written $a + (-b)$, the distributive property can be applied in the example below.

Example: Use the distributive property to write an equivalent algebraic expression.

$7(5m - 8)$

$7[5m + (-8)]$ *Rewrite subtraction as addition of* -8

$7(5m) + 7(-8)$ *Distributive property*

$35m - 56$ *Simplify*

In the following example, note that the factor of 2 is written to the right of the parentheses but is still distributed as before.

Example: Use the distributive property to write an equivalent algebraic expression:

$(3m + 4x - 10)2$

$(3m)2 + (4x)2 + (-10)2$ *Distributive property*

$6m + 8x - 20$ *Simplify*

In the following example, the negative sign in front of the parentheses can be interpreted as –1.

Example: Use the distributive property to write an equivalent algebraic expression:

$-(-2m + 6x)$

$-1(-2m + 6x)$

$-1(-2m) + (-1)(6x)$ *Distributive property*

$2m - 6x$ *Simplify*

Factoring

Factorization is the process of breaking up a mathematical quantity, such as a number or polynomial, into a product of two or more factors. For example, a factorization of the number 16 is $16 = 8 \times 2$. If multiplied out, the factorization results in the original number. A **prime factorization** is a specific factorization when the number is factored completely using prime numbers only. For example, the prime factorization of 16 is:

$$16 = 2 \times 2 \times 2 \times 2$$

A factor tree can be used to find the prime factorization of any number. Within a factor tree, pairs of factors are found until no other factors can be used, as in the following factor tree of the number 84:

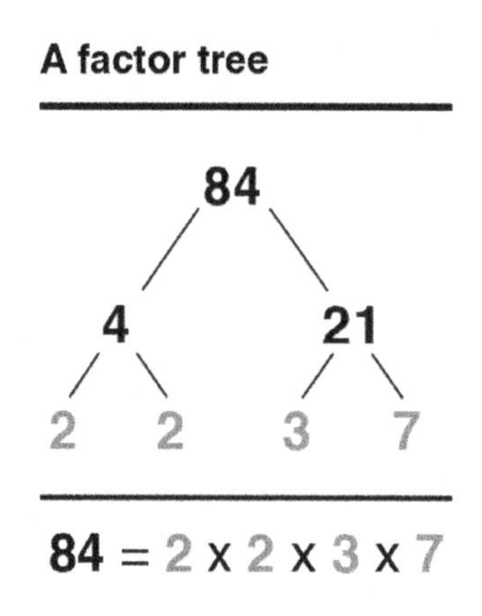

It first breaks 84 into 21×4, which is not a prime factorization. Then, both 21 and 4 are factored into their primes. Therefore,

$$84 = 2 \times 2 \times 3 \times 7$$

Factorization can be helpful in finding greatest common divisors and least common denominators.

Algebraic expression can also be factored. To factor a polynomial, first determine if there is a greatest common factor. If there is, factor it out. For example, $2x^2 + 8x$ has a greatest common factor of $2x$ and can be written as $2x(x + 4)$. After this, determine if the remaining polynomial follows a factoring pattern. If the polynomial has two terms, it could be a difference of squares, a sum of cubes, or a difference of cubes. If it falls into one of these categories, use the following rules:

$$a^2 - b^2 = (a + b)(a - b)$$

$$a^3 + b^3 = (a + b)(a^2 - ab + b^2)$$

$$a^3 - b^3 = (a - b)(a^2 + ab + b^2)$$

If there are three terms, and they follow one of the patterns below, then the expression can be factored as shown:

$$a^2 + 2ab + b^2 = (a + b)^2$$

$$a^2 - 2ab + b^2 = (a - b)^2$$

If not, try factoring into a product of two binomials in the form of $(x + p)(x + q)$. For example, to factor $x^2 + 6x + 8$, determine what two numbers have a product of 8 and a sum of 6. Those numbers are 4 and 2, so the trinomial factors into:

$$(x + 2)(x + 4)$$

Finally, if there are four terms, try factoring by grouping. Group together the first two terms, and then group together the last two terms. Identify the greatest common factor from each group (if there is one). Finally, factor out the common binomial factor of each expression. For example:

$$xy - x + 5y - 5$$

$$x(y - 1) + 5(y - 1)$$

$$(y - 1)(x + 5)$$

After the expression is completely factored, check the factorization by multiplying it out; if this results in the original expression, then the factoring is correct. Factorizations are helpful in solving equations in which a polynomial is set equal to 0. If the product of two algebraic expressions equals 0, then at least one of the factors must be equal to 0. Therefore, factor the polynomial within the equation, set each factor equal to 0, and solve. For example,

$$x^2 + 7x - 18 = 0$$

This can be solved by factoring into:

$$(x + 9)(x - 2) = 0$$

Set each factor equal to 0, and solve to obtain $x = -9$ and $x = 2$.

Identifying Zeros of Polynomials

A **polynomial function** is simply a function containing a polynomial expression. The *degree* of a polynomial depends on the largest exponent in the expression. Typical polynomial functions are **quartic,** with a degree of 4, **cubic,** with a degree of 3, or **quadratic,** with a degree of 2. Note that the exponents on the variables can only be nonnegative integers. An example of a quartic polynomial equation is:

$$y = x^4 + 3x^3 - 2x + 1$$

The "zeros" of a polynomial function are the points where its graph crosses the *y*-axis. In order to find the number of real zeros of a polynomial function, use **Descartes' Rule of Signs**, which states that the number of possible positive real zeros is equal to the number of sign changes in the coefficients. If there is only one sign change, there is only one positive real zero. In the example above, the signs of the coefficients are positive, positive, negative, and positive. The sign changes twice; therefore, there are at most two positive real zeros. The number of possible negative real zeros is equal to the number of sign changes in the coefficients when plugging $-x$ into the equation. Again, if there is only one sign change, there is only one negative real zero. The polynomial result when plugging -*x* into the equation is:

$$y = (-x)^4 + 3(-x)^3 - 2(-x) + 1$$

$$y = x^4 - 3x^3 + 2x + 1$$

The sign changes two times, so there are at most two negative real zeros.

Once the zeros are identified, they can be used to help sketch a graph of the polynomial.

Solving Linear Equations in One Variable

An **equation in one variable** is a mathematical statement where two algebraic expressions with the same variable (usually x) are set equal. To solve the equation, the variable must be isolated on one side of the equals sign. The addition and multiplication principles of equality are used to isolate the variable. The **addition principle of equality** states that the same number can be added to or subtracted from both sides of an equation. Because the same value is being applied to both sides, equality is maintained. For example, the equation $2x - 3 = 5x$ is equivalent to both

$$2x - 3 + 2 = 5x + 2 \text{ and } 2x - 3 - 5 = 5x - 5$$

This principle can be used to solve the following equation: $x + 5 = 4$. To isolate the variable x, subtract 5 from both sides of the equals sign.

Therefore:

$$x + 5 - 5 = 4 - 5$$

So, the solution is $x = -1$.

The **multiplication principle of equality** states that equality is maintained when both sides of an equation are multiplied or divided by the same number. For example, $4x = 5$ is equivalent to both $16x = 20$ and $x = \frac{5}{4}$. Multiplying both sides times 4 and dividing both sides by 4 maintains equality. Solving the equation $6x - 18 = 5$ requires the use of both principles. First, apply the addition principle to add 18 to both sides of the equation, which results in $6x = 23$. Then use the multiplication principle to divide both sides by 6, giving the solution $x = \frac{23}{6}$.

When solving linear equations, check the answer by plugging the solution back into the original equation. If the result is a false statement, something was done incorrectly during the solution procedure. Checking the example above gives the following:

$$6 \times \frac{23}{6} - 18 = 5$$

$$23 - 18 = 5$$

Therefore, the solution is correct.

Some equations in one variable involve fractions or require the use of the distributive property. In either case, the goal is to obtain only one variable term and then use the addition and multiplication principles to isolate that variable. Consider the equation $\frac{2}{3}x = 6$. To solve for x, multiply each side of the equation by the reciprocal of $\frac{2}{3}$, which is $\frac{3}{2}$. This step results in:

$$\frac{3}{2} \times \frac{2}{3}x = \frac{3}{2} \times 6$$

This simplifies into the solution $x = 9$. Now consider the equation:

$$3(x + 2) - 5x = 4x + 1$$

Use the distributive property to clear the parentheses. Therefore, multiply each term inside the parentheses by 3. This step results in:

$$3x + 6 - 5x = 4x + 1$$

Collecting like terms on the left-hand side results in:

$$-2x + 6 = 4x + 1$$

Finally, apply the addition and multiplication principles. Add $2x$ to both sides to obtain:

$$6 = 6x + 1$$

Then, subtract 1 from both sides to obtain $5 = 6x$. Finally, divide both sides by 6 to obtain the solution $\frac{5}{6} = x$.

Two other types of solutions can be obtained when solving an equation in one variable. There could be no solution, or the solution set could contain all real numbers. Consider the equation:

$$4x = 6x + 5 - 2x$$

First, combine the like terms on the right to obtain:

$$4x = 4x + 5$$

Next, subtract $4x$ from both sides. This step results in the false statement $0 = 5$. There is no value that can be plugged into x that will make this equation true. Therefore, there is no solution. The solution procedure contained correct steps, but the result of a false statement means that no value satisfies the equation. The symbolic way to denote that no solution exists is $\emptyset$.

Next, consider the equation:

$$5x + 4 + 2x = 9 + 7x - 5$$

Combining the like terms on both sides results in:

$$7x + 4 = 7x + 4$$

The left-hand side is exactly the same as the right-hand side. Using the addition principle to move terms, the result is $0 = 0$, which is always true. Therefore, the original equation is true for any number, and the solution set is all real numbers. The symbolic way to denote such a solution set is $\mathbb{R}$, or in interval notation, $(-\infty, \infty)$.

To re-state it simply, a linear equation in one variable can be solved using the following steps:

1. Simplify both sides of the equation by removing all parentheses, using the distributive property, and collecting all like terms.

2. Collect all variable terms on one side of the equation and all constant terms on the other side by adding the same quantity to or subtracting the same quantity from both sides.

3. Isolate the variable by either multiplying or dividing both sides of the equation by the same number.

4. Check the answer.

If an equation contains multiple fractions, it might make sense to clear the equation of fractions first by multiplying all terms by the least common denominator. Also, if an equation contains several decimals, it might make sense to clear the decimals as well by multiplying times a factor of 10. If the equation has decimals in the hundredth place, multiply every term in the equation by 100.

Solving Linear Inequalities in One Variable

A linear equation in one variable can be written in the form $ax + b = 0$. A **linear inequality** is very similar, although the equals sign is replaced by an inequality symbol such as $<, >, \leq$, or $\geq$. Some examples of linear inequalities in one variable are $2x + 3 < 0$ and $4x - 2 \leq 0$. Solving an inequality involves finding the set of numbers that, when plugged into the variable, make the inequality a true statement. These numbers are known as the **solution set** of the inequality. To solve an inequality, use the same properties that are necessary when solving equations. First, add or subtract variable terms and/or constants to isolate all variable terms on one side of the equals sign and all constant terms on the other side. Then, either multiply or divide both sides by the same number to obtain an inequality that gives the solution set. When multiplying or dividing b a negative number, change the direction of the inequality symbol. Consider the linear inequality:

$$-2x - 5 > x + 6$$

First, add 5 to both sides and subtract x from both sides to obtain $-3x > 11$. Then, divide both sides by -3, making sure to change the direction of the inequality symbol. These steps result in the solution $x < -\frac{11}{3}$. Therefore, any number less than $-\frac{11}{3}$ satisfies this inequality.

Solving Quadratic Equations in One Variable

A **quadratic equation** in standard form can have either two solutions, one solution, or two complex solutions (no real solutions).

$$ax^2 + bx + c = 0$$

To figure out how many solutions a quadratic equation will have, use the *determinant*: $b^2 - 4ac$. If the determinant is positive, there are two real solutions. If the determinant is negative, there are no real solutions. If the determinant is equal to 0, there is one real solution. For example, given the quadratic equation:

$$4x^2 - 2x + 1 = 0$$

Its determinant is:

$$(-2)^2 - 4(4)(1) = 4 - 16 = -12$$

Therefore, this equation has two complex solutions, which also means that it has no real solutions.

There are quite a few ways to solve a quadratic equation. The first is by **factoring**. If the equation is in standard form and the polynomial can be factored, set each factor equal to 0. For example:

$$x^2 - 4x + 3 = (x - 3)(x - 1)$$

The solutions to the following equation are those that satisfy both $x - 3 = 0$ and $x - 1 = 0$, or $x = 3$ and $x = 1$:

$$x^2 - 4x + 3 = 0$$

This is the simplest method to solve quadratic equations; however, not all quadratic equations can be factored.

Another method is **completing the square**. If a polynomial in a quadratic equation cannot be factored, completing the square means adding a constant to make it a perfect square. For example, the polynomial in the following equation cannot be factored, so the next step is to complete the square to find its solutions.

$$x^2 + 10x - 9 = 0$$

First, move the constant to the right side by adding 9 to both sides, which results in:

$$x^2 + 10x = 9$$

Then divide the coefficient of x by 2, square it, and add the result to both sides of the equation. In this example, $\left(\frac{10}{2}\right)^2 = 25$ is added to both sides of the equation to obtain:

$$x^2 + 10x + 25 = 9 + 25$$

The left-hand side can then be factored into $(x + 5)^2 = 34$. Solving for x then involves taking the square root of both sides and subtracting 5. This leads to the two solutions:

$$x = \pm\sqrt{34} - 5$$

The third method is the **quadratic formula.** Given a quadratic equation in standard form, $ax^2 + bx + c = 0$, its solutions always can be found using the formula:

$$x = \frac{-b \pm \sqrt{b^2 - 4ac}}{2a}$$

Equivalent Expressions Involving Rational Exponents and Radicals

Re-writing complex radical expressions as equivalent forms with rational exponents can help to simplify them. The rule that helps this conversion is:

$$\sqrt[n]{x^m} = x^{\frac{m}{n}}$$

If $m = 1$, the rule is simply:

$$\sqrt[n]{x} = x^{\frac{1}{n}}$$

For instance, consider the following expression:

$$\sqrt[4]{x}\,\sqrt[2]{y}$$

This can be written as one radical expression, but first it needs to be converted to an equivalent expression. The equivalent expression is $x^{\frac{1}{4}}y^{\frac{1}{2}}$. The goal is to have one radical, which means one index *n*, so a common denominator of the exponents must be found. The common denominator is 4, so an equivalent expression is:

$$x^{\frac{1}{4}}y^{\frac{2}{4}}$$

The exponential rule can be used to factor $\frac{1}{4}$ out of both variables:

$$a^m b^m = (ab)^m$$

This process results in the expression $(xy^2)^{\frac{1}{4}}$, and its equivalent radical form is:

$$\sqrt[4]{xy^2}$$

Another type of problem could involve going in the opposite direction—starting with rational exponents and using an equivalent radical form to simplify the expression. For instance, $32^{\frac{1}{5}}$ might not appear equal to 2, but putting it in its equivalent radical form $\sqrt[5]{32}$ shows that it is the fifth root of 32, which is 2.

Solving Systems of Equations

Linear Inequalities in Two Variables

A system of linear inequalities *in two variables* consists of two inequalities with two variables, *x* and *y*. For example, the following is a system of linear inequalities in two variables:

$$\begin{cases} 4x + 2y < 1 \\ 2x - y \leq 0 \end{cases}$$

The brace on the left side shows that the two inequalities are grouped together. A solution of a single inequality in two variables is an ordered pair that satisfies the inequality. For example, (1, 3) is a solution of the linear inequality $y \geq x + 1$ because when plugged in, it results in a true statement. The graph of an inequality in two variables consists of all ordered pairs that make the solution true. Therefore, the entire solution set of a single inequality contains many ordered pairs, and the set can be graphed by using a half plane. A **half plane** consists of the set of all points on one side of a line. If the inequality consists of $>$ or $<$, the line is dashed because no solutions actually exist on the line shown. If the inequality consists of $\geq$ or $\leq$, the line is solid, and solutions are on the line shown. To graph a linear inequality, graph the corresponding equation found by replacing the inequality symbol with an equals sign. Then pick a test point that exists on either side of the line. If that point results in a true statement when plugged into the original inequality, shade in the side containing the test point. If it results in a false statement, shade in the opposite side.

Solving a system of linear inequalities must be done graphically. Follow the process as described above for both given inequalities. The solution set to the entire system is the region that is in common to every graph in the system.

For example, here is the solution to the following system:

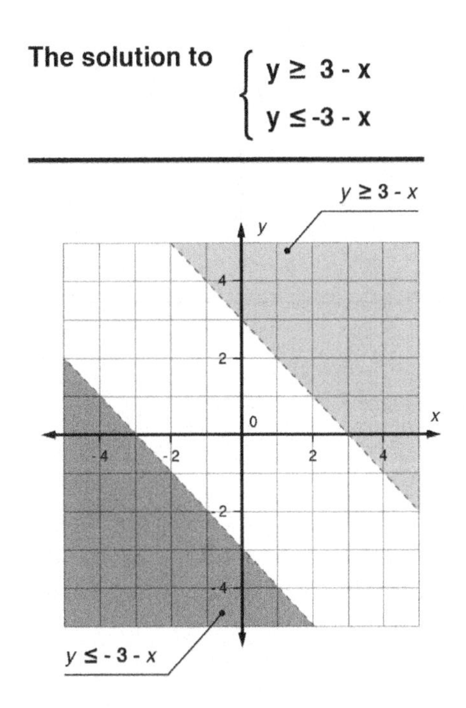

Note that there is no region in common, so this system has no solution.

Creating, Solving, or Interpreting Systems of Two Linear Equations in Two Variables

An example of a system of two linear equations in two variables is the following:

$$2x + 5y = 8$$

$$5x + 48y = 9$$

A solution to a system of two linear equations is an ordered pair that satisfies both equations in the system. A system can have one solution, no solution, or infinitely many solutions. The solution can be found through a **graphing** technique. The solution to a system of equations is actually equal to the point where both lines intersect. If the lines intersect at one point, there is one solution, and the system is *consistent*. If the two lines are parallel, they will never intersect, so there is no solution. In this case, the system is *inconsistent.* Third, if the two lines are actually the same line, there are infinitely many solutions, and the solution set is equal to the entire line. These lines are *dependent.* Here is a summary of the three cases:

Solving Systems by Graphing

Consistent	Inconsistent	Dependent
One solution	No solution	Infinite number of solutions
Lines intersect	*Lines are parallel*	*Coincide: same line*

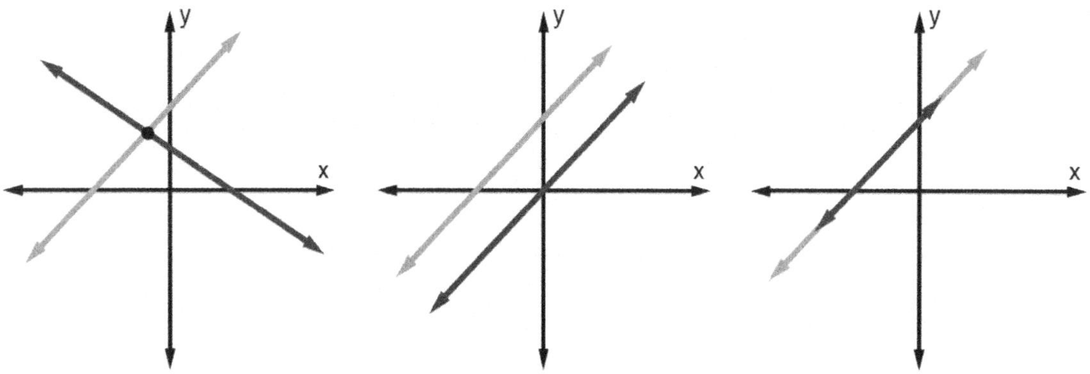

Consider the following system of equations:

$$y + x = 3$$

$$y - x = 1$$

To find the solution graphically, graph both lines on the same *xy*-plane. Graph each line using either a table of ordered pairs, the *x*- and *y*-intercepts, or slope and the *y*-intercept. Then, locate the point of intersection.

The graph is shown here:

The System of Equations $\begin{cases} y + x = 3 \\ y - x = 1 \end{cases}$

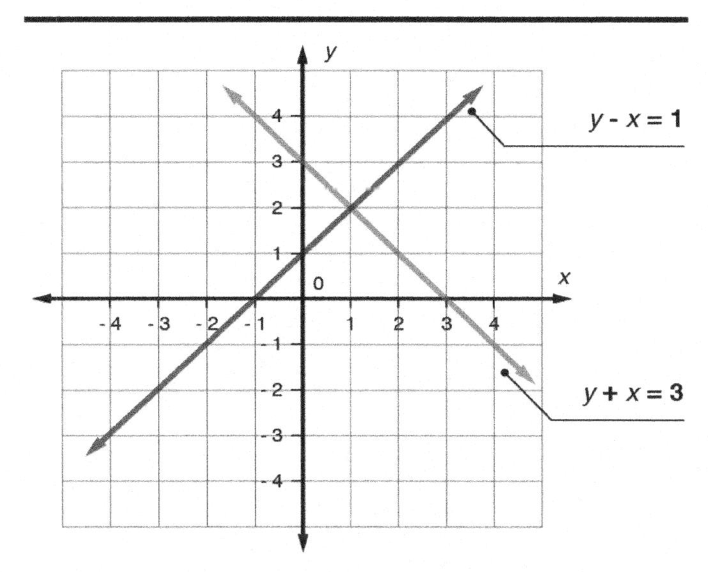

The point of intersection is the ordered pair (1, 2). This solution can be checked by plugging it back into both original equations to make sure it results in true statements:

$$2 + 1 = 3$$
$$2 - 1 = 1$$

Both equations are true, so the solution is correct.

The following system has no solution:

$$y = 4x + 1$$
$$y = 4x - 1$$

Both lines have the same slope and different y-intercepts, so they are parallel, meaning that they run alongside each other and never intersect.

Finally, the following solution has an infinite number of solutions:

$$2x - 7y = 12$$
$$4x - 14y = 24$$

Note that the second equation is equal to the first equation times 2. Therefore, they are the same line. The solution can be written in set notation as:

$$\{(x, y) | 2x - 7y = 12\}$$

This represents the entire line.

Linear Equations in Two Variables

There are two algebraic methods for finding solutions to systems of linear equations. The first is **substitution**. This process is better suited for systems when one of the equations is already solved for one variable, or when solving for one variable is easy to do. The equation that is already solved for one variable is substituted into the other equation, which results in a linear equation in one variable. This equation can be solved for the given variable, and then that solution can be plugged into one of the original equations, which can then be solved for the other variable. This last step is known as **back-substitution**, and the end result is an ordered pair.

The following is an example of a system that is suited for substitution:

$$y = 4x + 2$$

$$2x + 3y = 9$$

In the second equation, $4x + 2$ can be substituted for y, resulting in:

$$2x + 3(4x + 2) = 9$$

Distributing and solving the second equation looks like this:

$$2x + 12x + 6 = 9$$

$$14x + 6 = 9$$

$$14x = 3$$

$$x = \frac{3}{14}$$

Now, the solution for x can plug into the first equation to solve for y:

$$y = 4(\frac{3}{14}) + 2$$

$$y = 2\frac{6}{7} \ or \ \frac{20}{7}$$

The other method is known as **elimination,** or the **addition method**. This is better suited for equations in the standard form of:

$$Ax + By = C$$

The goal in this method is to multiply one or both equations by numbers that result in opposite coefficients. Then, add the equations together to obtain an equation in one variable. Solve for the given variable, then take that value and back-substitute to obtain the other part of the ordered pair solution.

The following is an example of a system that is suited for elimination:

$$2x + 3y = 8$$

$$4x - 2y = 10$$

Start by multiplying the first equation by -2 so that the x variable will be eliminated when the equations are added together:

$$-4x - 6y = -16$$

Then add the two equations to get $-8y = -6$. Solving for y results in $y = \frac{3}{4}$. Input the solved value of y into the first equation to solve for x:

$$2x + 3(\frac{3}{4}) = 8$$

$$2x + \frac{9}{4} = 8$$

$$2x = \frac{23}{4}$$

$$x = \frac{23}{8}$$

In order to check an answer when solving a system of equations, the solutions must be plugged into both original equations to show that it solves not only one of the equations, but both of them.

If either solution results in an untrue statement when inserted into the original equation, then there is no solution to the system.

Graphing Functions

Functions can be displayed graphically to analyze behaviors and patterns within data. Different functions have certain characteristics that make their corresponding graphs distinctive.

Polynomial Functions

A polynomial equation is a polynomial set equal to another polynomial, or in standard form, a polynomial is set equal to 0. A **polynomial function** is a polynomial set equal to *y*. For instance, the following is a polynomial:

$$x^2 + 2x - 8$$

And this is a polynomial equation:

$$x^2 + 2x - 8 = 0$$

The following is the corresponding polynomial function:

$$y = x^2 + 2x - 8$$

To solve a polynomial equation, the *x*-values in which the graph of the corresponding polynomial function crosses the *x*-axis are sought. These coordinates are known as the *zeros* of the polynomial function because they are the coordinates in which the *y*-coordinates are 0. One way to find the zeros of a polynomial is to find its factors, then set each individual factor equal to 0, and solve each equation to find the zeros. A **factor** is a linear expression, and to completely factor a polynomial, the polynomial must be rewritten as a product of individual linear factors. The polynomial listed above can be factored as $(x + 4)(x - 2)$. Setting each factor equal to zero results in the zeros $x = -4$ and $x = 2$.

Here is the graph of the zeros of the polynomial:

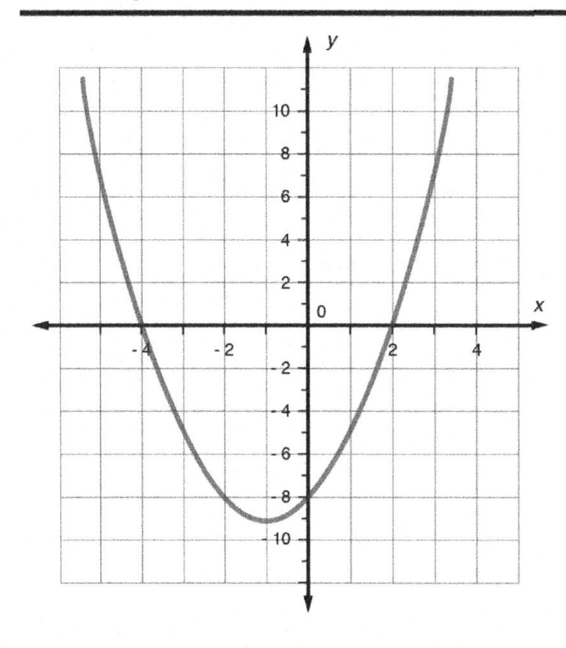

Nonlinear Relationships

A common nonlinear relationship between two variables involves inverse variation where one quantity varies inversely with respect to another. The equation for inverse variation is $y = \frac{k}{x}$, where k is still known as the constant of variation. Here is the graph of the curve $y = \frac{3}{x}$:

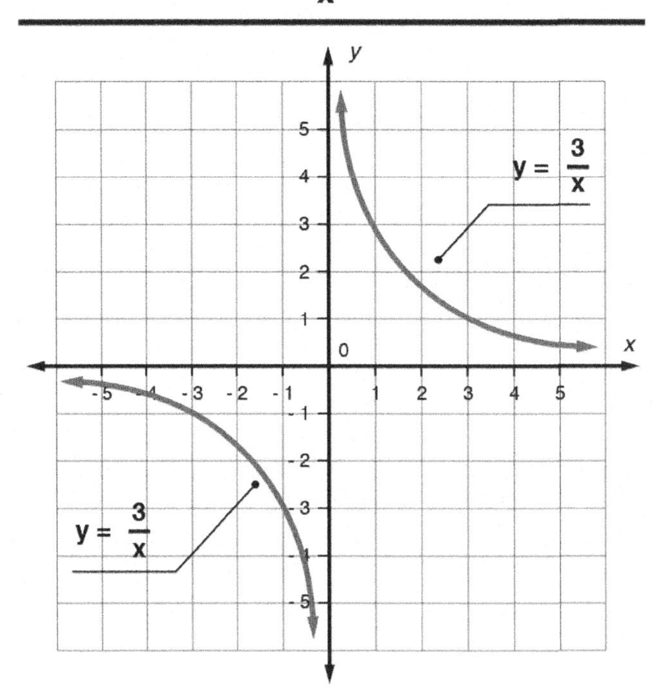

Another common nonlinear function is the squaring function $f(x) = x^2$, seen here:

f(x) = x²

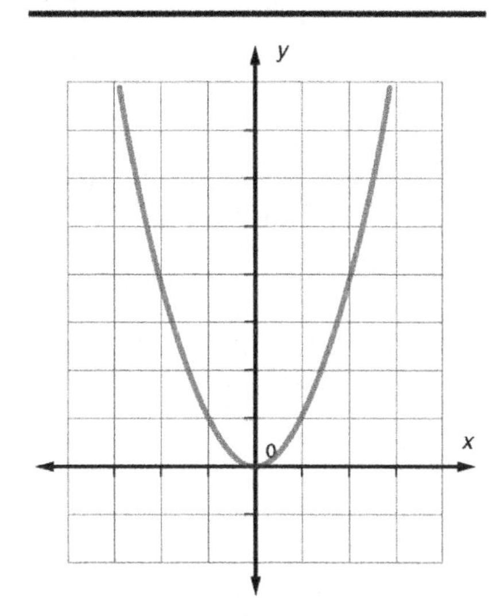

Notice that as the independent variable x increases when $x > 0$, the dependent variable y also increases. However, y does not increase at a constant rate.

Root Functions

A **radical expression** is an expression involving a root (square root, cube root, etc.). A radical function is a function that involves a radical expression. For example, $\sqrt{x+1}$ is a radical expression, and the corresponding function is:

$$y = \sqrt{x+1}$$

It can also be written in function notation as:

$$f(x) = \sqrt{x+1}$$

If the root is even, the radicand must be positive. Therefore, in order to find the domain of a radical function with an even index, set the radicand greater than or equal to zero and find the set of numbers that satisfies that inequality.

$$f(x) = \sqrt{x+1}$$

The domain of the above is all numbers greater than or equal to -1. The range of this function is all nonnegative real numbers because the square root, or any even root, can never output a negative number. The domain of an odd root is all real numbers because the radicand can be negative in an odd root. Take a look at this square root function:

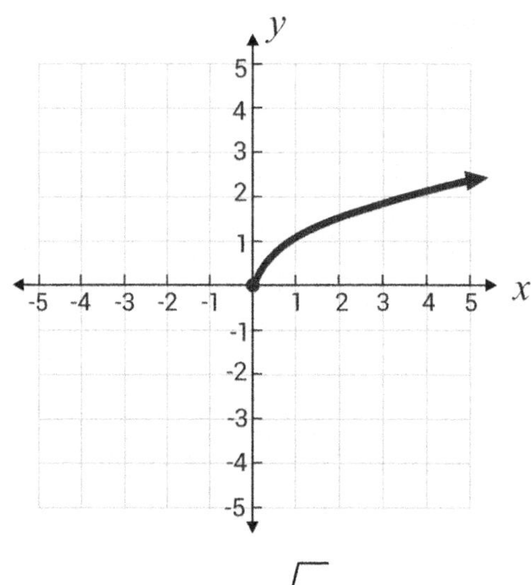

$$y = \sqrt{x}$$

Here is a cube root function:

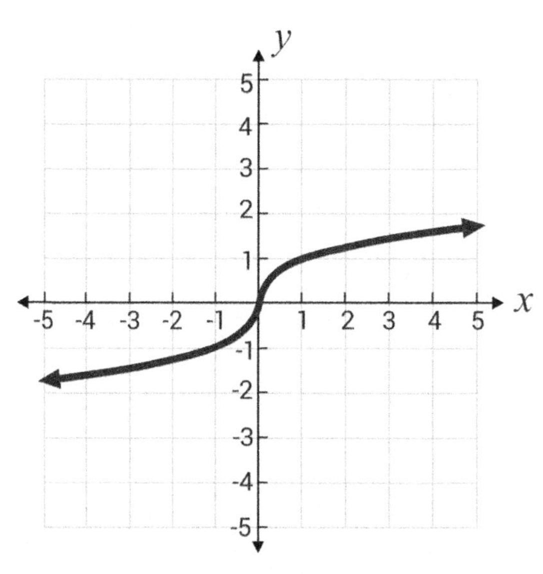

$$y = \sqrt[3]{x}$$

Absolute Value Functions

Absolute value functions contain an expression that uses absolute value symbols. The vertex is (0, 0), and the axis of symmetry divides the graph on the y-axis as seen below:

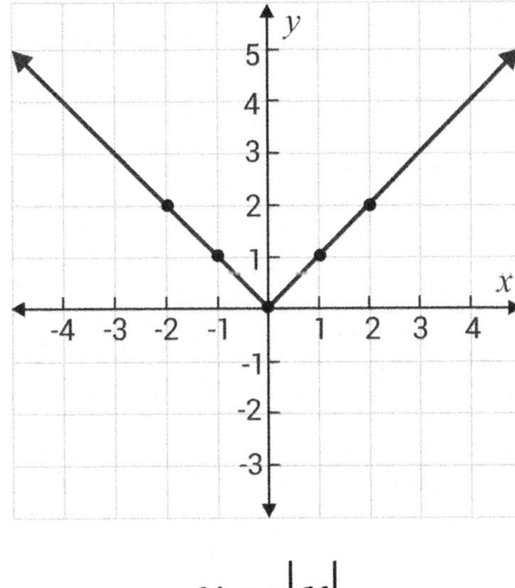

$$y = |x|$$

Piecewise Functions

A **piecewise function** is a function that is defined in pieces or sections. The graph of the function behaves differently over different intervals along the *x*-axis, or different intervals of its domain. Therefore, the function is defined using different mathematical expressions over these intervals. The function is not defined by only one equation. In a piecewise function, the function is actually defined by two or more equations, where each equation is used over a specific interval of the domain.

Here is a graph of a piecewise function:

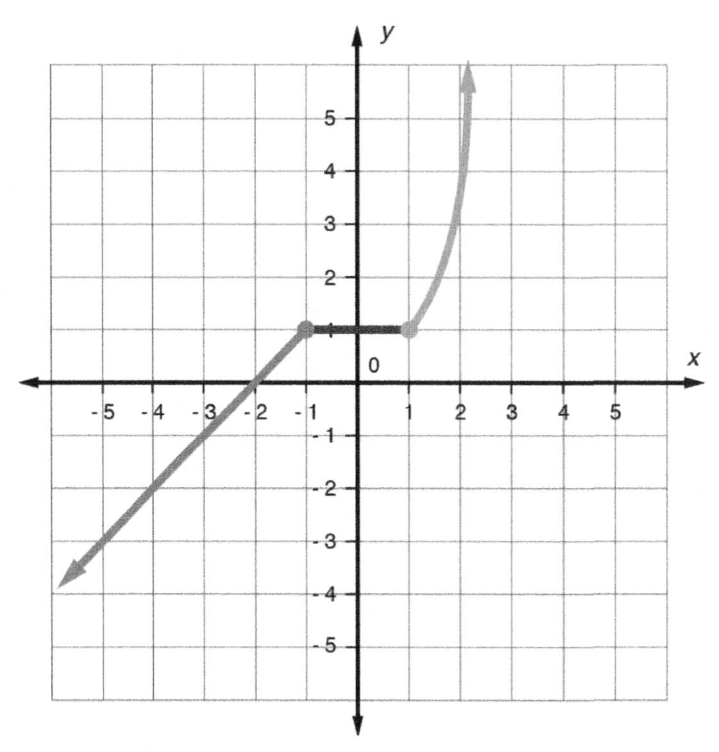

Notice that from $(-\infty, -1]$, the graph is a line with positive slope. From $[-1, 1]$ the graph is a horizontal line. Finally, from $[1, \infty)$ the graph is a nonlinear curve. Both the domain and range of this function are all real numbers, expressed as $(-\infty, \infty)$.

Piecewise functions can also have discontinuities or jumps in the graph. Here is the graph of a piecewise function with discontinuities at $x = 1$ and $x = 2$:

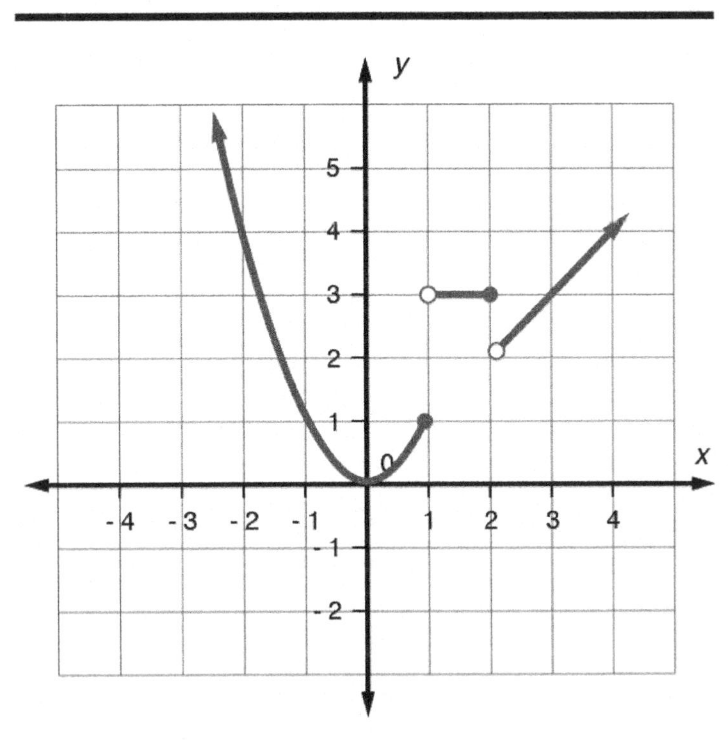

The open circle at a point indicates that the endpoint is not included in that part of the graph, and the closed circle indicates that the endpoint is included. The domain of this function is all real numbers; however, the range is all non-negative real numbers $[0, \infty)$.

Translating Phrases and Sentences into Expressions, Equations, and Inequalities

When presented with a real-world problem, the first step is to determine what unknown quantity must be solved for. Use a variable, such as x, to represent that unknown quantity. Sometimes there can be two or more unknown quantities. In this case, either choose an additional variable, or if a relationship exists between the unknown quantities, express the other quantities in terms of the original variable. After choosing the variables, form algebraic expressions and/or equations that represent the verbal statement in the problem.

The following table shows examples of vocabulary used to represent the different operations:

Addition	Sum, plus, total, increase, more than, combined, in all
Subtraction	Difference, less than, subtract, reduce, decrease, fewer, remain
Multiplication	Product, multiply, times, part of, twice, triple
Division	Quotient, divide, split, each, equal parts, per, average, shared

Inequalities can also exist within verbal mathematical statements. These expressions state that given quantities are *less than* ($<$), *less than or equal to* ($\leq$) , *greater than* ($>$), or *greater than or equal to* ($\geq$). Another type of inequality is when a quantity is said to be *not equal to* another quantity ($\neq$).

Interpreting Linear Functions

A linear function of the form $f(x) = mx + b$ has two important quantities: m and b. The quantity m represents the **slope** of the line, and the quantity b represents the **y-intercept** of the line. When the function represents a real-life situation or a mathematical model, these two quantities are very meaningful. The slope, m, represents the rate of change, or the amount y increases/decreases given an increase in x. If m is positive, the rate of change is positive, and if m is negative, the rate of change is negative. The y-intercept, b, represents the amount of quantity y when x is 0. In many applications, if the x-variable is never a negative quantity, the y-intercept represents the initial amount of the quantity y. The x-variable often represents time, so it makes sense that it would not be negative.

Consider the following situation. A taxicab driver charges a flat fee of $2 per ride and $3 a mile. This statement can be modeled by the function:

$$f(x) = 3x + 2$$

Where x represents the number of miles and $f(x) = y$ represents the total cost of the ride. The total cost increases at a constant rate of $2 per mile, and that is why this situation is a linear relationship. The slope $m = 3$ is equivalent to this rate of change. The flat fee of $2 is the y-intercept. It is the place where the graph crosses the x-axis, and it represents the cost when $x = 0$, or when no miles have been traveled in the cab. The y-intercept in this situation represents the flat fee.

Consider another example. These two equations represent the cost (C) of t-shirts (x) at two different printing companies:

$$C(x) = 7x$$

$$C(x) = 5x + 25$$

The first equation represents a scenario in which each t-shirt costs $7. In this equation, x varies directly with y. There is no y-intercept, which means that there is no initial cost for using that printing company. The rate of change is 7, which is price per shirt. The second equation represents a scenario that has both an initial cost and a cost per t-shirt. The slope of 5 shows that each shirt is $5. The y-intercept of 25 shows that there is an initial cost of using that company. Therefore, it makes sense to use the first company at $7 per shirt when only purchasing a small number of t-shirts. However, any large orders would be cheaper from the second company because eventually that initial cost would become negligible.

Functions and Function Notation

A **relation** is any set of ordered pairs (x, y). The values listed first in the ordered pairs, known as the x-coordinates, make up the **domain** of the relation. The values listed second, known as the y-coordinates, make up the **range**. A relation in which every member of the domain corresponds to only one member of the range is known as a **function.** Functions are most often given in terms of equations instead of ordered pairs. For instance, here is an equation of a line:

$$y = 2x + 4$$

In function notation, this can be written as:

$$f(x) = 2x + 4$$

The expression $f(x)$ is read "f of x," where f stands for *function*. When different x-values get plugged into the function, the output is $y = f(x)$. The set of all inputs are in the domain and the set of all outputs are in the range.

The x-values are known as the **independent variables** of the function, and the y-values are known as the **dependent variables** of the function. The y-values depend on the x-values. For instance, if $x = 2$ is plugged into the function shown above, the y-value depends on that input.

$$f(2) = 2 \times 2 + 4 = 8.$$

Therefore, $f(2) = 8$, which is the same as writing the ordered pair (2, 8). To graph a function, graph it in equation form. Therefore, replace $f(x)$ with y and plot ordered pairs.

Here is an example of a relation that is not a function:

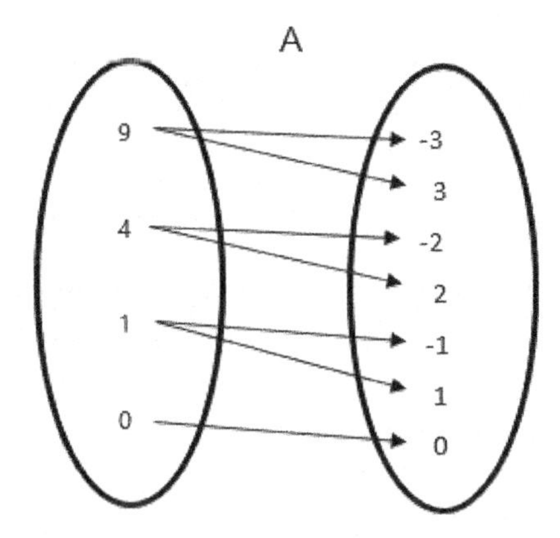

Every member of the first set, the domain, corresponds to two members of the second set, the range. Therefore, this relation is not a function.

A function can also be represented by a table of ordered pairs, a graph of ordered pairs (a scatterplot), or a set of ordered pairs as shown in the following:

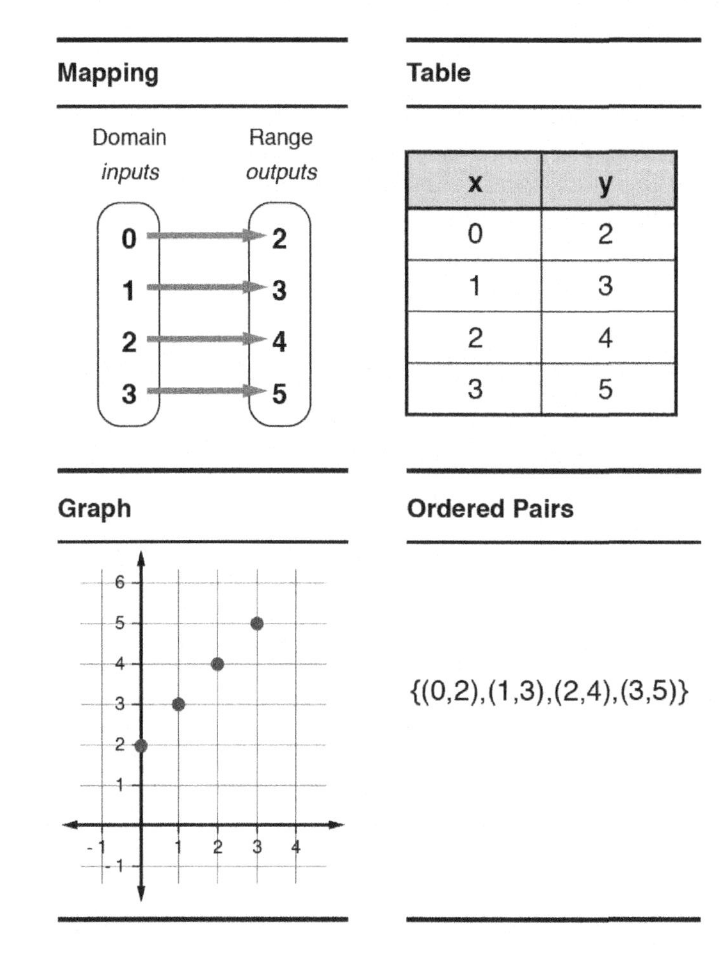

Table

x	y
0	2
1	3
2	4
3	5

{(0,2),(1,3),(2,4),(3,5)}

The relation shown is a function because every member of the domain corresponds to exactly one member of the range.

The following graph shows a linear function because the relationship between the two variables is constant. Each time the distance increases by 25 miles, 1 hour passes. This pattern continues for the rest of the graph. The line represents a constant rate of 25 miles per hour. This graph can also be used to solve problems involving predictions for a future time. After 8 hours of travel, the rate can be used to predict the distance covered. The equation at the top of the graph corresponds to this rate.

For a time of 10 hours, the distance would be 250 miles, as the equation yields:

$$d = 25 \times 10 = 250\ miles$$

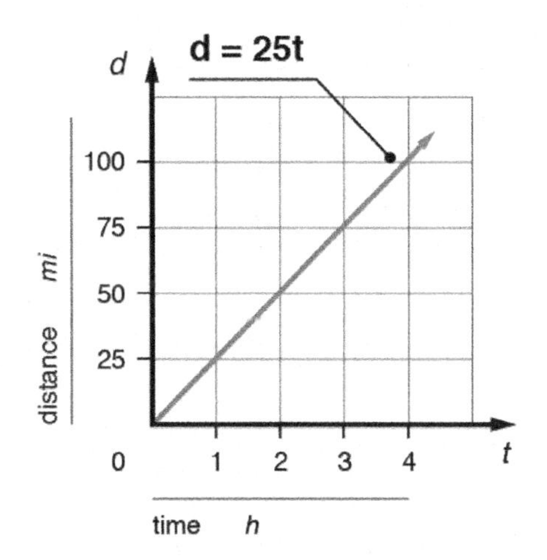

Another representation of a linear relationship can be seen in a table. In the table below, the y-values increase by 3 as the x-values increase by 1. This pattern shows that the relationship is linear. The y-value can each time be determined by multiplying the x-value by 3, then adding 1. The following equation models this relationship:

$$y = 3x + 1$$

y = 3x + 1	
x	y
0	1
1	4
2	7
4	13
5	16

Exponential growth involves the dependent variable changing by a common ratio every unit increase. The equation of exponential growth is $y = a^x$ for $a > 0, a \neq 1$. The value a is known as the **base.** Consider the exponential equation $y = 2^x$. When $x = 1$, $y = 2$, and when $x = 2$, $y = 4$. For every unit increase in x, the value of the output variable doubles. In the following graph of $y = 2^x$, notice that as the dependent variable (y) gets very large, x increases slightly.

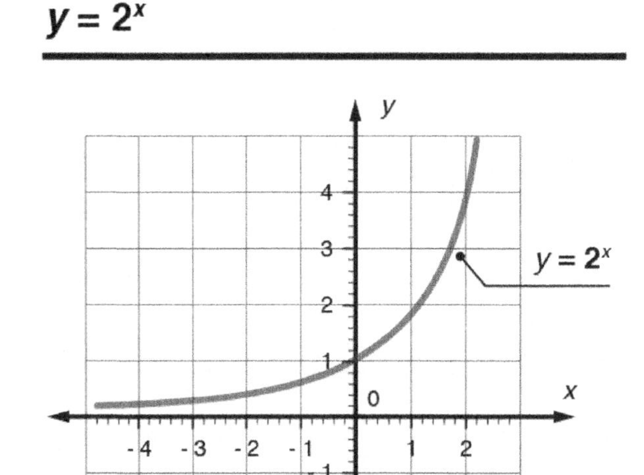

According to the definition of a function, there can only be one value of y for each value of x. Therefore, all graphs of functions must pass the **vertical line test**: if any vertical line intersects a graph in more than one place, the graph is not that of a function. For instance, the graph of a circle is not a function because one can draw a vertical line through a circle and intersect the circle twice. Common functions include lines and polynomials, which pass the vertical line test.

Key features of functions and their graphs include the following:

Increasing/Decreasing Intervals: Intervals where the x-values of a function are increasing or decreasing

Absolute maximum/minimum: The highest and lowest y-value in a function

Relative maximum/minimum: Turning points where the graph changes from increasing to decreasing or from decreasing to increasing

Positive/Negative: A function is positive when the function rises above the x-axis and negative when the function falls below the x-axis.

Symmetry: Symmetry occurs when there is a line (axis of symmetry) that divides the function in half so that both sides are the same.

End Behavior: What happens to the graph as it extends in either direction

Periodicity: Values in a function repeating in a regular pattern

Writing a Function that Describes a Relationship Between Two Quantities

Functions can describe relationships that exist within numerical patterns. For example, the following table shows two sets of numbers that each follow their own pattern.

The first column shows a pattern of numbers increasing by 1. The second column shows the numbers increasing by 4. The numbers in the first column correspond to those in the second. A question to ask is, "How can the number in the first column turn into the number in the second column?"

1	4
2	8
3	12
4	16
5	20

This answer will lead to the relationship between the two sets. The first set of numbers is multiplied by 4 to get the second set. To confirm this relationship, check each pair of corresponding numbers. The function that represents this table is:

$$f(x) = 4x$$

In some cases, the relationship between sets of numbers is simply addition, subtraction, multiplication, or division. In other relationships, these operations are used in conjunction with each together. The relationship in the following table uses both multiplication and addition. The following expression shows this relationship: $3x + 2$. The x represents the numbers in the first column.

1	5
2	8
3	11
4	14

Explaining the Steps in Solving a Simple Equation

One-step problems take only one mathematical step to solve. For example, solving the equation $5x = 45$ is a one-step problem because the one step of dividing both sides of the equation by 5 is the only step necessary to obtain the solution $x = 9$. The multiplication principle of equality is the one step used to isolate the variable.

A multi-step problem involves more than one step to find the solution, or it could consist of solving more than one equation. An example of a two-step equation is

$$2x - 4 = 5$$

To solve, add 4 to both sides and then divide both sides by 2. This process gives the answer $\frac{9}{2}$.

An example of a problem involving two separate equations is:

$$y = 3x \text{ and } 2x + y = 4$$

The two equations form a system that must be solved together in two variables. The system can be solved by the substitution method. Since y is already solved for in terms of x, replace y with $3x$ in the equation:

$$2x + y = 4$$

This results in:

$$2x + 3x = 4$$

Therefore, $5x = 4$ and $x = \frac{4}{5}$. Because there are two variables, the solution requires a value for both x and y. Substitute $x = \frac{4}{5}$ into either original equation to find y. The easiest choice is $y = 3x$. Therefore,

$$y = 3 \times \frac{4}{5} = \frac{12}{5}$$

The solution can be written as the ordered pair $\left(\frac{4}{5}, \frac{12}{5}\right)$.

Real-world problems can be translated into both one-step and multi-step problems. In either case, the word problem must be translated from its verbal form into mathematical expressions and equations that can be solved using algebra. An example of a one-step real-world problem is the following: *A cat weighs half as much as a dog living in the same house. If the dog weighs 14.5 pounds, how much does the cat weigh?* To solve this problem, first put it into equation form and define variables that represent the unknown quantities. For this problem, let x be equal to the unknown weight of the cat. Because two times the weight of the cat equals 14.5 pounds, the equation to be solved is $2x = 14.5$. Use the multiplication principle to divide both sides by 2. Therefore, $x = 7.25$, and the cat weighs 7.25 pounds.

Most of the time, real-world problems require multiple steps. The following is an example of a multi-step problem: *The sum of two consecutive page numbers is equal to 437. What are those page numbers?* First, define the unknown quantities. If x is equal to the first page number, then $x + 1$ is equal to the next page number because they are consecutive integers. Their sum is equal to 437. Putting this information together results in the equation:

$$x + x + 1 = 437$$

To solve, first collect like terms to obtain:

$$2x + 1 = 437$$

Then, subtract 1 from both sides and divide by 2. The solution to the equation is $x = 218$. Therefore, the two consecutive page numbers that satisfy the problem are 218 and 219. It is always important to make sure that answers to real-world problems make sense. For instance, if the solution to this same problem resulted in decimals, that should be a red flag. Page numbers are whole numbers; therefore, if decimals are found to be answers, the solution process should be double-checked for mistakes.

Geometry and Measurement

Transformations in the Plane

Two-dimensional figures can undergo various types of transformations in a geometric plane. They can be shifted horizontally and vertically, reflected, compressed, or stretched.

A **shift**, also known as a slide or a translation, moves the shape in one direction. Here is an example:

A Translation

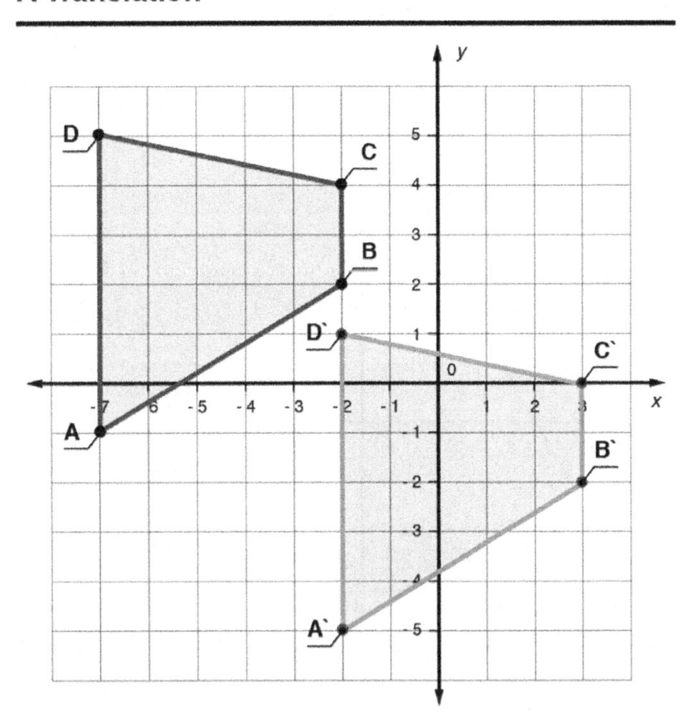

Notice that the size of the original shape has not changed at all. If the shift occurs within a **Cartesian coordinate system**, the standard x and y coordinate plane, it can be represented by adding to or subtracting from the x and y coordinates of the original shape. All vertices will move the same number of units because the shape and size of the shape do not change.

A figure can also be **reflected,** or flipped, over a given line known as the **line of reflection**. For instance, consider the following image:

A Reflection Over the Y-Axis

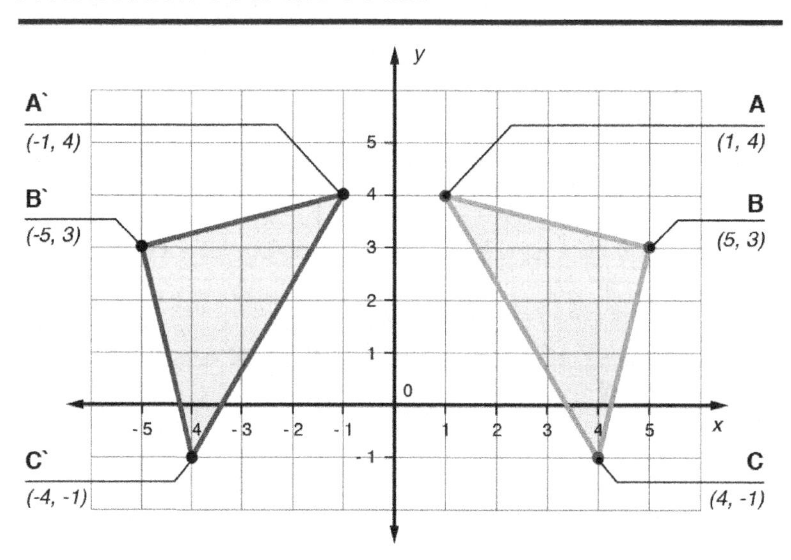

After the reflection, the original shape remains the same size, but the coordinates change. In a reflection across the *y*-axis, as shown above, the *y*-coordinates stay the same, but the *x*-coordinates become negative. For example, the triangle above starts in the right-hand quadrants and is reflected over the *y*-axis to the left-hand quadrants. Point A has the initial coordinates of (1, 4), but in the reflection, the point A' becomes (-1, 4).

Similarly, if the shape is reflected over the *x*-axis, the *x*-coordinate stays the same, but the *y*-coordinates become negative. For instance, in the image below, the point *C* at (3, 5) becomes *C'* at (3, -5).

A Reflection Over the X-Axis

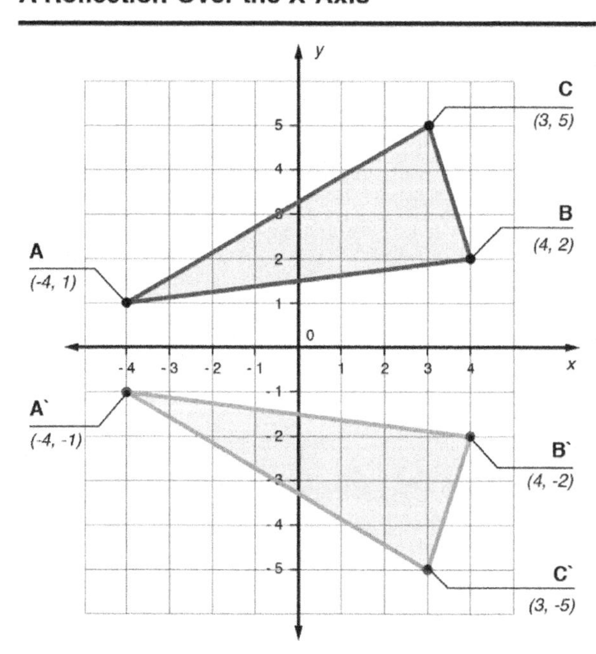

A **compression** or **stretch** of a figure involves changing the size of the original figure; both transformations are called **dilations**. A compression shrinks the size of the figure, while a stretch results in a larger figure. Here is an image of a square stretched to double its original size.

A Dilation with a Scale Factor of 2

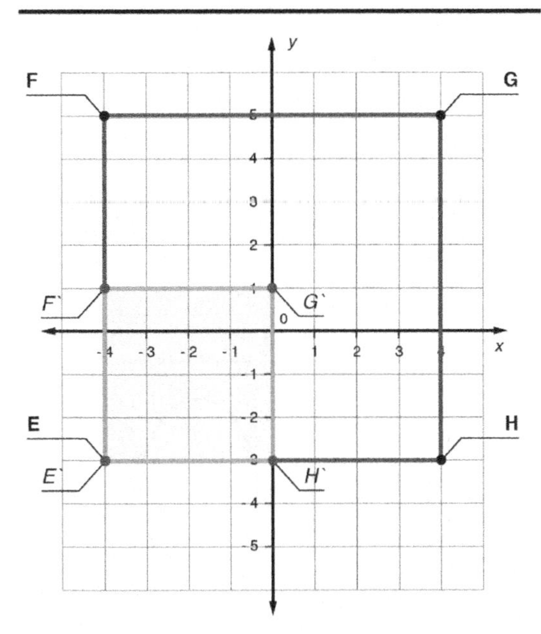

If a shape within the Cartesian coordinate system gets stretched, its coordinates get multiplied by a number greater than 1, and if a shape gets compressed, its coordinates get multiplied by a number between 0 and 1.

A figure can undergo any combination of transformations. For instance, it can be shifted, reflected, and stretched at the same time.

<u>*Properties of Polygons and Circles*</u>

Shapes are defined by their angles and number of sides. A shape with one continuous side, where all points on that side are equidistant from a center point is called a **circle.** A shape made with three straight line segments is a **triangle.** A shape with four sides is called a **quadrilateral,** but more specifically a **square**, **rectangle, parallelogram**, or **trapezoid,** depending on the interior angles. These shapes are two-dimensional and only made of straight lines and angles.

Solids can be formed by combining these shapes and forming three-dimensional figures. While two-dimensional figures have only length and height, three-dimensional figures also have depth. Examples of solids may be prisms or spheres.

The four figures below have different names based on their sides and dimensions. *Figure 1* is a **cone**, a three-dimensional solid formed by a circle at its base and the sides combining to one point at the top. *Figure 2* is a **triangle**, a shape with two dimensions and three line segments. *Figure 3* is a **cylinder** made up of two base circles and a rectangle to connect them in three dimensions. *Figure 4* is an **oval** formed by one continuous line in two dimensions; it differs from a circle because not all points are equidistant from the center.

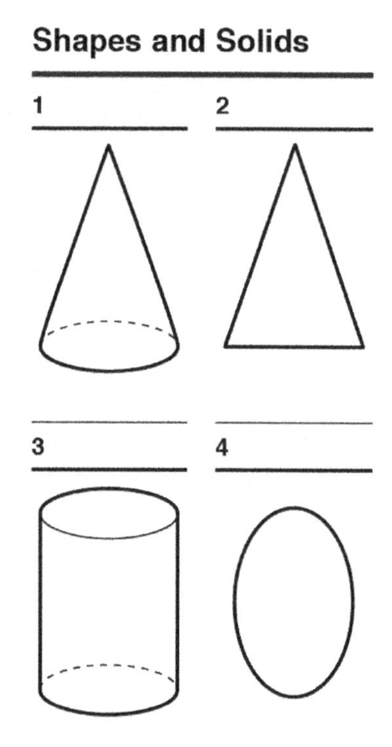

The **cube** in *Figure 5* below is a three-dimensional solid made up of squares. *Figure 6* is a **rectangle** because it has four sides that intersect at right angles. More specifically, it is a **square** because the four sides are equal in length.

Figure 7 is a **pyramid** because the bottom shape is a square and the sides are all triangles. These triangles intersect at a point above the square. *Figure 8* is a **circle** because it is made up of one continuous line where the points are all equidistant from one center point.

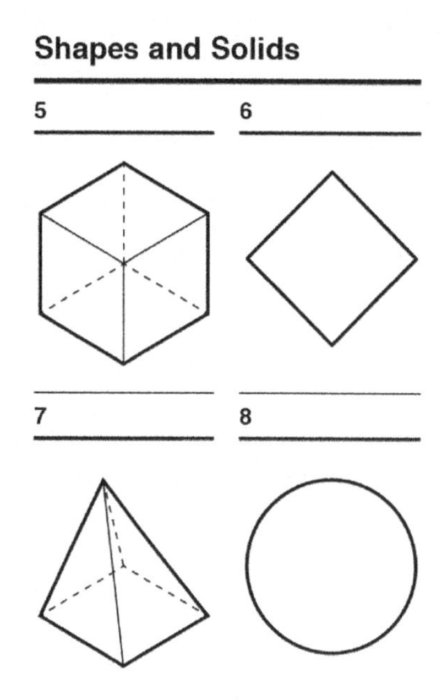

Perimeter and Area

Perimeter and area are geometric quantities that describe objects' measurements. **Perimeter** is the distance around an object. The perimeter of an object can be found by adding the lengths of all sides. Perimeter may be useful in problems dealing with lengths around objects such as fences or borders. It may also be useful in finding missing lengths. If the perimeter is given, but a length is missing, subtraction can be used to find the missing length. Given a square with side length s, the formula for perimeter is $P = 4s$. Given a rectangle with length l and width w, the formula for perimeter is:

$$P = 2l + 2w$$

The perimeter of a triangle is found by adding the three side lengths, and the perimeter of a trapezoid is found by adding the four side lengths. The units for perimeter are always the original units of length, such as meters, inches, miles, etc. When discussing a circle, the distance around the object is referred to as its **circumference,** not perimeter. The formula for the circumference of a circle is $C = 2\pi r$, where *r* represents the radius of the circle, which is the distance between the center of the circle and any point along its circumference. The symbol π represents the constant *pi,* which is usually abbreviated as 3.14. This formula can also be written as $C = d\pi$, where *d* represents the diameter of the circle, which is the distance of a line segment that cuts the circle directly in half.

Area is the two-dimensional space covered by an object. Finding the area may require a simple formula or multiple formulas used together.

The units for area are square units, such as square meters, square inches, and square miles. Given a square with side length s, the formula for its area is $A = s^2$.

The table below shows some other common shapes and their area formulas:

Shape	Formula	Graphic
Rectangle	$Area = length \times width$	
Triangle	$Area = \frac{1}{2} \times base \times height$	height base
Circle	$Area = \pi \times radius^2$	radius

The following formula for the area of a trapezoid is not as widely used as those shown in the table, but it is still very important:

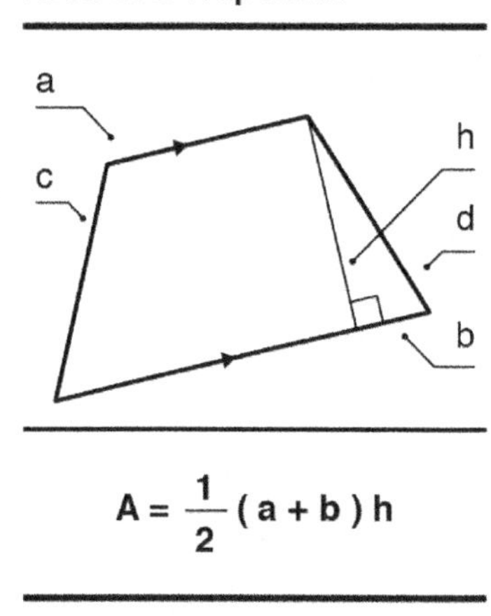

Complex shapes might require more than one formula. To find the area of the figure below, break the figure into two shapes. The rectangle's dimensions are 6 cm by 7 cm. The triangle has a base of 4 cm and a height of 6 cm. Plug the rectangle's dimensions into the proper formula:

$$A = 6 \times 7$$

Multiplication yields an area of 42 cm^2. The triangle's area can be found using the formula:

$$A = \frac{1}{2} \times 4 \times 6$$

Multiplication yields an area of 12 cm^2. Add the two areas to find the total area of the figure, which is 54 cm^2.

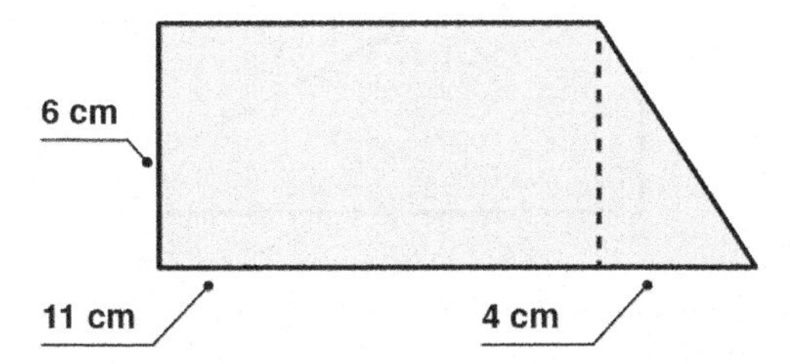

The previous problem required adding or combining areas, but some problems may require subtracting them. To find the area of the shaded region in the figure below, determine the area of the whole figure. Then the area of the circle can be subtracted from the whole.

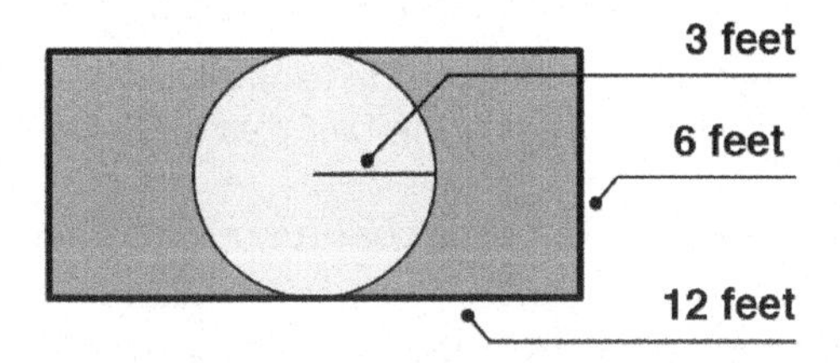

The following formula shows the area of the outside rectangle:

$$A = 12 \times 6 = 72\ ft^2$$

The area of the inside circle can be found by the following formula:

$$A = \pi(3)^2 = 9\pi = 28.3\ ft^2$$

To find the shaded region's area, subtract the area of the circle from the area of the rectangle to yield $43.7\ ft^2$.

Geometric figures may be shown as pictures or described in words, as in the following problem. *What is the perimeter of a rectangular playing field that is 95 meters long and 50 meters wide?* Identifying the perimeter requires finding the distance around the field, including two lengths and two widths. This quantity can be calculated using the following equation:

$$P = 2(95) + 2(50) = 290\ m$$

Therefore, the distance around the field is 290 meters.

The Pythagorean Theorem

The **Pythagorean Theorem** expresses an important relationship between the three sides of a right triangle. It states that the square of the hypotenuse is equal to the sum of the squares of the other two sides. When using the Pythagorean Theorem, the **hypotenuse** is the longest side of the triangle.

The theorem can be seen in the following diagram:

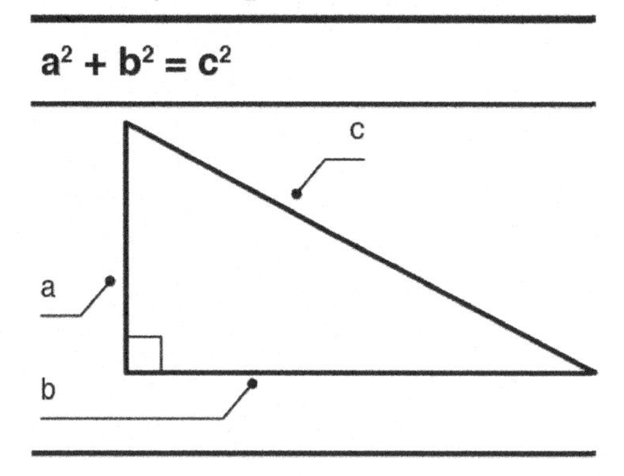

The Pythagorean Theorem can be used to find a missing side of a right triangle.

Using Congruence and Similarity Criteria for Triangles

Two figures are **congruent** if they have the same shape and same size, meaning equal angle measurements and side lengths. Two figures are **similar** if they have the same angle measurements but not side lengths. If two pairs of angles are congruent in two triangles, then those triangles are similar because their third angles have to be equal due to the fact that all three angles add up to 180 degrees.

There are five main theorems that are used to prove congruence in triangles. Each theorem involves showing that different combinations of sides and angles are the same in two triangles, which proves the triangles are congruent. The **side-side-side (SSS) theorem** states that if all sides are equal in two triangles, the triangles are congruent. The **side-angle-side (SAS) theorem** states that if two pairs of sides and the included angles are equal in two triangles, then the triangles are congruent. Similarly, the **angle-side-angle (ASA) theorem** states that if two pairs of angles and the included side lengths are equal in two triangles, the triangles are similar. The **angle-angle-side (AAS) theorem** states that two triangles are congruent if they have two pairs of congruent angles and a pair of corresponding equal side lengths that are not included. Finally, the **hypotenuse-leg (HL) theorem** states that if two right triangles have equal hypotenuses and an equal pair of shorter sides, the triangles are congruent. An important item to note is that **angle-angle-angle (AAA)** is not enough information to prove congruence because the three angles could be equal in two triangles, but their sides could be different lengths.

Using Volume Formulas to Solve Problems

Perimeter and area are two-dimensional descriptions; volume is three-dimensional. **Volume** describes the amount of space that an object occupies, but it differs from area because it has three dimensions instead of two. The units for volume are **cubic units**, such as cubic meters, cubic inches, and cubic miles. Volume can be found with formulas for common objects such as cylinders and boxes.

The following chart shows a formula and diagram for the volume of two objects:

Shape	Formula	Diagram
Rectangular Prism (box)	$V = length \times width \times height$	length height width
Cylinder	$V = \pi \times radius^2 \times height$	radius height

Volume formulas of these two objects are derived by finding the area of the bottom two-dimensional shape, such as the circle or rectangle, and then multiplying times the height of the three-dimensional shape.

Other volume formulas include the volume of a cube with side length:

$$V = s^3$$

the volume of a sphere with radius:

$$V = \frac{4}{3}\pi r^3$$

and the volume of a cone with radius and height:

$$V = \frac{1}{3}\pi r^2 h$$

If a soda can has a height of 5 inches and a radius on the top of 1.5 inches, the volume can be found using one of the given formulas. A soda can is a cylinder. Knowing the given dimensions, the formula can be completed as follows:

$$V = \pi\ (radius)^2 \times height = \pi\ (1.5)^2 \times 5 = 35.325\ inches^3$$

Use the volume formula for rectangular prisms to calculate the volume of the following image:

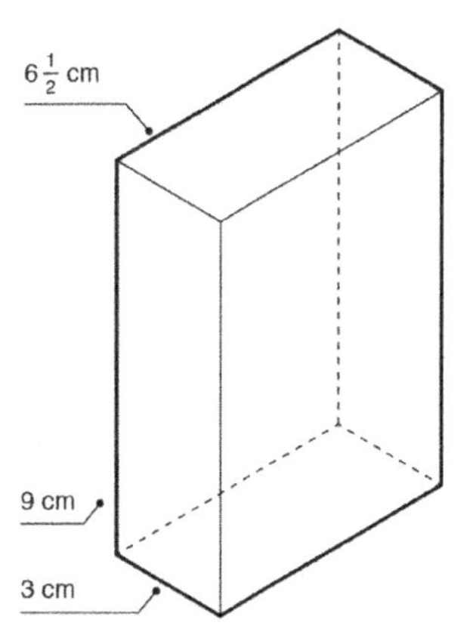

Inputting the measurements into the volume formula results in the following equation:

$$V = 6\frac{1}{2} \times 3 \times 9$$

When dealing with fractional edge lengths, it is helpful to convert to improper fractions. The length 6 ½ cm becomes $\frac{13}{2}$ cm. Then the formula becomes:

$$V = \frac{13}{2} \times 3 \times 9$$

$$\frac{13}{2} \times \frac{3}{1} \times \frac{9}{1} = \frac{351}{2}$$

This value for volume is better understood when turned into a mixed number, which would be 175 ½ cm^3.

When dimensions for length are given with fractional parts, it can be helpful to turn the mixed number into an improper fraction, multiply to find the volume, then convert back to a mixed number. When finding surface area, the conversion to improper fractions can also be helpful. **Surface area**, or the sum of the areas of a solid's different sides, can be found for the same prism above by breaking down the figure into its basic shapes. These shapes are rectangles, made up of the two bases, two sides, and the front and back. The formula for surface area adds the areas for each of these shapes in the following equation:

$$SA = \left(6\frac{1}{2} \times 3\right) + \left(6\frac{1}{2} \times 3\right) + (3 \times 9) + (3 \times 9) + (6\frac{1}{2} \times 9) + (6\frac{1}{2} \times 9)$$

Because there are so many terms in a surface area formula and because this formula contains a fraction, it can be simplified by combining groups that are the same.

The formula can be simplified to:

$$SA = 2\left(6\frac{1}{2} \times 3\right) + 2(3 \times 9) + 2\left(6\frac{1}{2} \times 9\right)$$

$$2\left(\frac{13}{2} \times 3\right) + 2(27) + 2\left(\frac{13}{2} \times 9\right)$$

$$2\left(\frac{39}{2}\right) + 54 + 2\left(\frac{117}{2}\right)$$

$$39 + 54 + 117 = 210 \text{ cm}^2$$

Determining How Changes to Dimensions Change Area and Volume

When the dimensions of an object change, the area and volume are also subject to change. For example, the following rectangle has an area of 98 square centimeters:

$$A = 7 \times 14 = 98\text{cm}^2$$

If the length is increased by 2, becoming 16 cm, then the area becomes:

$$A = 7 \times 16 = 112\text{cm}^2$$

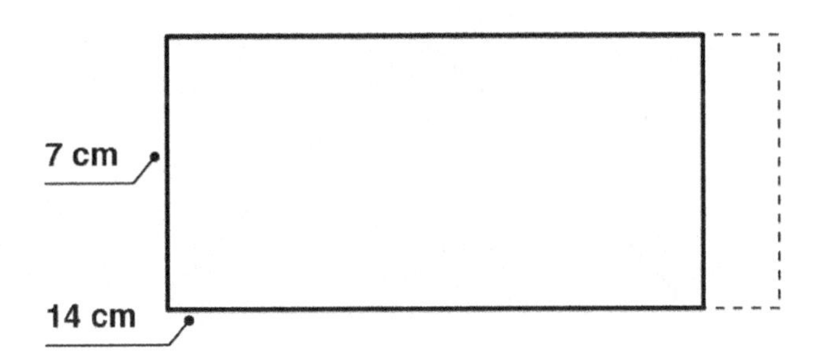

For the volume of an object, there are three dimensions. The given prism has a volume of:

$$V = 4 \times 12 \times 3 = 144\text{m}^3$$

If the height increases by 3, the volume becomes:

$$V = 4 \times 12 \times 6 = 288\text{m}^3$$

When the height increased by 3, it doubled, which also resulted in the volume doubling. If the original width had doubled from 4 to 8 cm, the volume would be:

$$V = 8 \times 12 \times 3 = 288\text{m}^3$$

When the width doubled, the volume doubled also. The same increase in volume would result if the length was doubled.

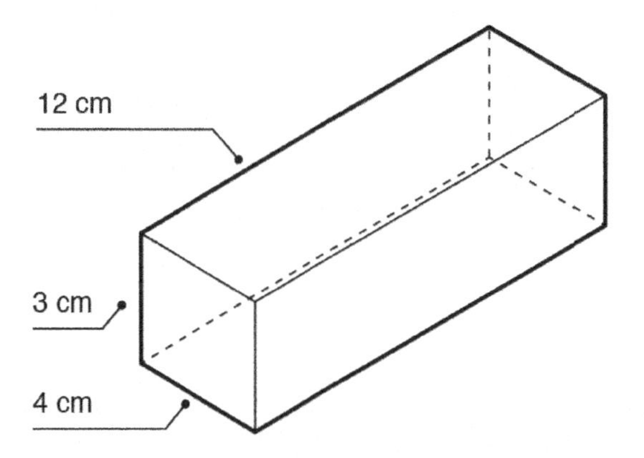

Figure Nets

The **net of a figure** is the resulting two-dimensional shapes when a three-dimensional shape is broken down. For example, the net of a cone is shown below. The base of the cone is a circle, shown at the bottom. The rest of the cone is a quarter of a circle. If the cone is cut down the side and laid out flat, these would be the resulting shapes:

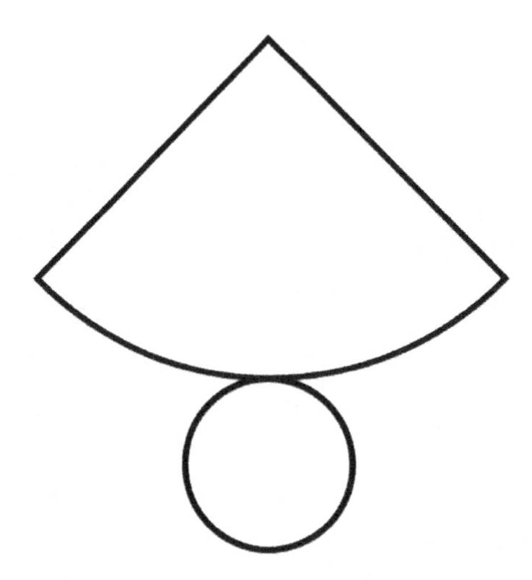

A net for a pyramid is shown in the figure below. The base of the pyramid is a square. The four shapes coming off the square are triangles. Bringing the triangles together at the top results in a pyramid:

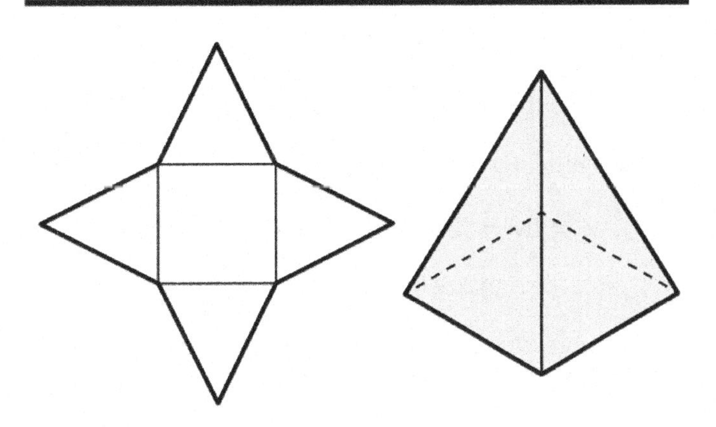

The net of a cylinder is shown below. When the cylinder is broken down, the bases are circles, and the side is a rectangle wrapped around the circles. The circumference of the circle turns into the length of the rectangle:

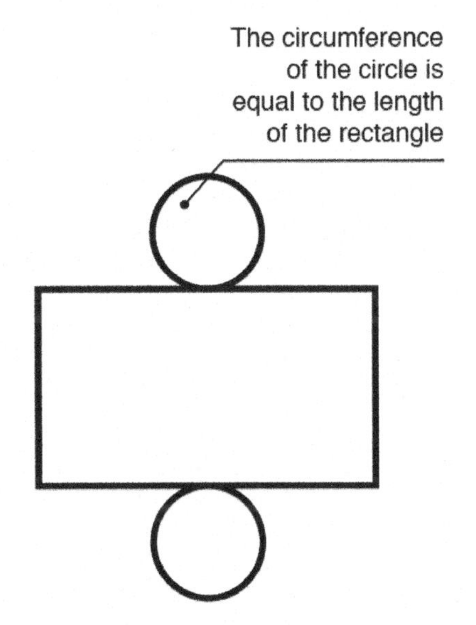

Nets can be used when calculating different values for given shapes. One useful way to calculate surface area is to find the net of the object, then find the areas each shape and add them together. Nets are also useful when composing or decomposing shapes and when determining connections between objects.

Using Nets to Determine the Surface Areas of Three-Dimensional Figures

The surface area of a three-dimensional figure is the total area of each of the figure's faces. Because nets lay out each face of an object, they make it easier to visualize and measure surface area. The image on the following page shows the dimensions for each face of the triangular prism. To determine the area for the two triangles, use the following formula:

$$A = \frac{1}{2}bh$$

$$\frac{1}{2} \times 8 \times 9 = 36cm^2$$

Since there are two triangles, double this measurement to get 72cm^2.

The rectangles' areas can be described by the equation:

$$A = lw = (8 \times 5) + (9 \times 5) + (10 \times 5)$$

$$40 + 45 + 50 = 135\text{cm}^2$$

Adding the areas of both triangles and the rectangles yields a total surface area of 207cm^2.

A Triangular Prism and Its Net

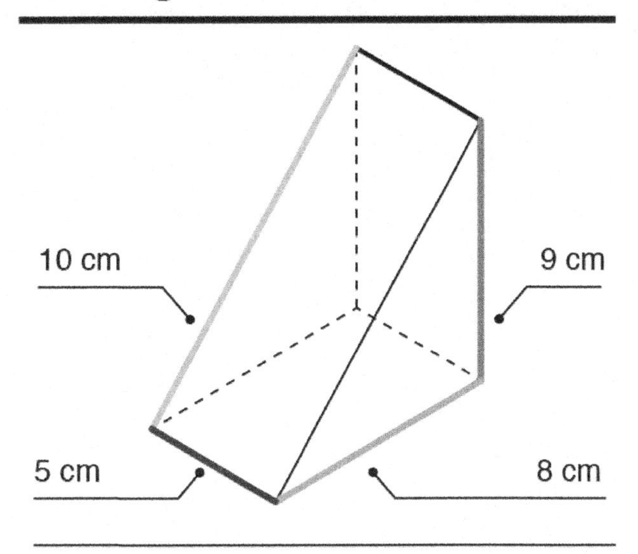

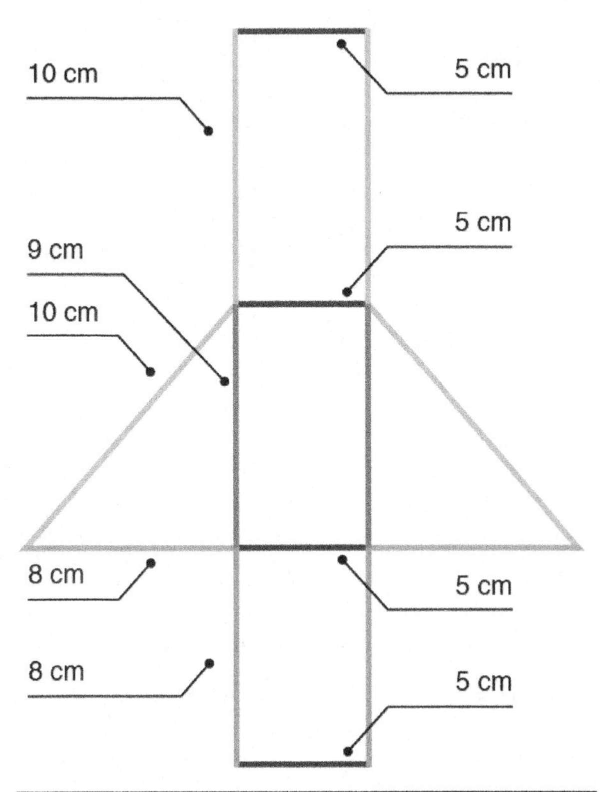

$$\text{SA} = 2 \times \left(\frac{1}{2}\text{bh}\right) + \text{lw}$$

$$= 2 \times \left(\frac{1}{2} \times 8 \times 9\right) + (8 \times 5 + 9 \times 5 + 10 \times 5)$$

$$= 207\text{cm}^2$$

The following picture shows the net for a rectangular prism and the dimensions for each shape making up the prism. The surface area is the sum of each rectangle added together. Calculations for the prism's surface area are shown below the net.

A Rectangular Prism and its Net

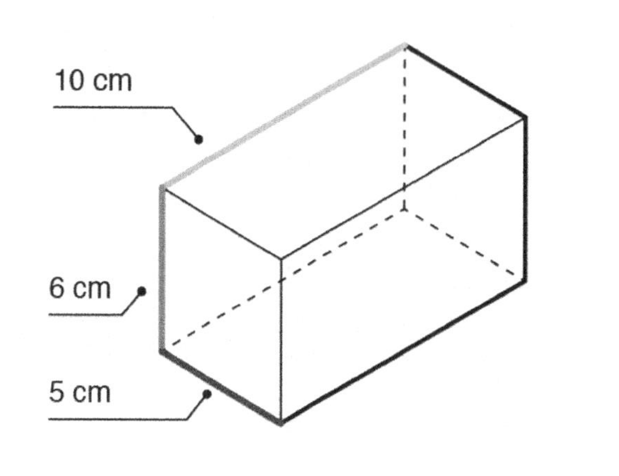

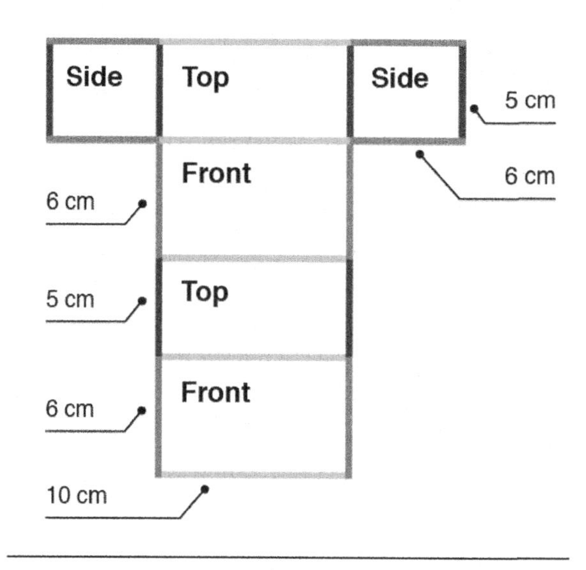

SA = 5×10 + 5×6 + 6×10 + 5×6 + 5×10 + 6×10

= 50 + 30 + 60 + 30 + 50 + 60

= 280cm^2

Line Segments, Rays, and Lines

A **line segment** is made up of two connected endpoints. A **ray** is made up of one endpoint and one extending side that goes on forever. A **line** has no endpoints and two sides that extend forever.

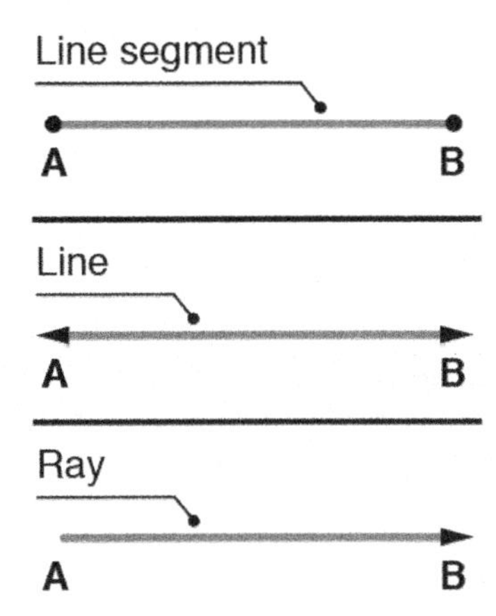

A set of lines can be parallel, perpendicular, or neither, depending on how the two lines interact. **Parallel** lines run alongside each other but never intersect. **Perpendicular** lines intersect at a 90-degree (*right*) angle. The image below shows both an example and a non-example of each set of lines. Because the first set of lines, in the top left corner, will eventually intersect if they continue, they are not parallel. In the second set, the lines run in the same direction and will never intersect, making them parallel. The lines in the third set, in the bottom left corner, intersect at an angle that is not 90 degrees. The lines in the fourth set are perpendicular because they intersect at exactly a right angle.

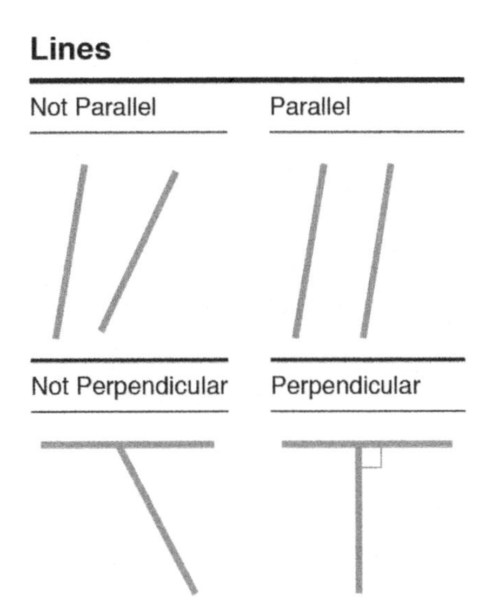

Classifying Angles Based on Their Measure

When two rays join together at their endpoints, they form an **angle**. An angle whose measure is ninety degrees is a right angle. Ninety degrees is a standard to which other angles are compared. If an angle is less than ninety degrees, it is an **acute angle**. If it is greater than ninety degrees, it is an **obtuse angle**. If an angle is equal to twice a right angle, or 180 degrees, it is a **straight angle**.

Examples of each type of angle are shown below:

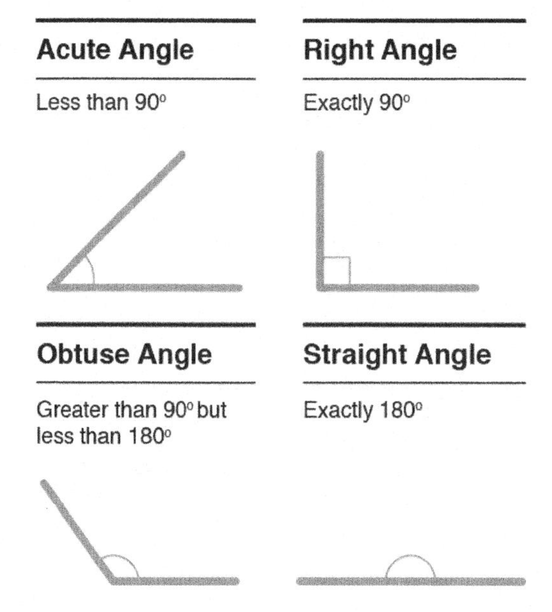

Solving for Missing Values in Triangles, Circles, and Other Figures

The sum of the measures of any triangle's three angles is 180 degrees. Therefore, if only two angles are known, the third can be found by subtracting the sum of the two known quantities from 180. Two angles that add up to 90 degrees are known as **complementary angles.** For example, angles measuring 72 and 18 degrees are complementary. Finally, two angles that add up to 180 degrees are known as **supplementary angles.** To find the supplement of an angle, subtract the given angle from 180 degrees. For example, the supplement of an angle that is 50 degrees is:

$$180 - 50 = 130 \text{ degrees}$$

Consider the following problem: The measure of an angle is 60 degrees less than two times the measure of its complement. What is the angle's measure? To solve this, let x be the unknown angle. Therefore, its complement is $90 - x$. The problem gives that:

$$x = 2(90 - x) - 60$$

To solve for x, distribute the 2, and collect like terms. This process results in:

$$x = 120 - 2x$$

Then, use the addition property to add $2x$ to both sides to obtain $3x = 120$. Finally, use the multiplication properties of equality to divide both sides by 3 to get $x = 40$. Therefore, the angle measures 40 degrees. Also, its complement measures 50 degrees.

For the triangle below, the one given angle has a measure of 55 degrees. The missing angle is x. The third angle is labeled with a square, which indicates a measure of 90 degrees. Because all angles must add up to 180 degrees, the following equation can be used to find the missing x-value:

$$55 + 90 + x = 180$$

Adding the two given angles and subtracting the total from 180 gives an answer of 35 degrees.

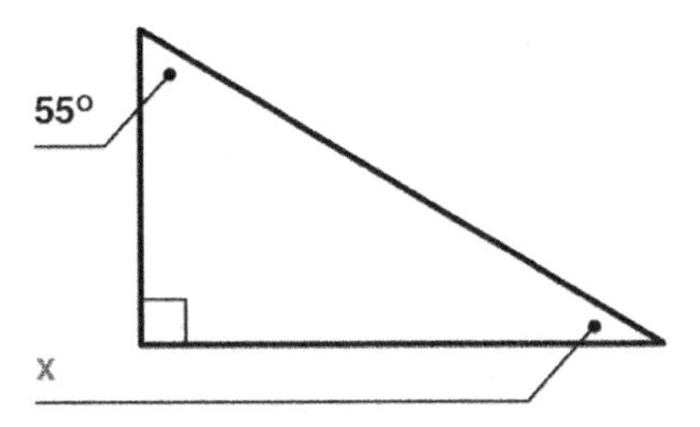

A similar problem can be solved with circles. If the radius is given, but the circumference is unknown, it can be calculated based on the formula $C = 2\pi r$. In the figure below, the radius of 8 cm can be substituted for r in the formula. Then the circumference can be found as:

$$C = 2\pi \times 8 = 16\pi \ = 50.24\ cm$$

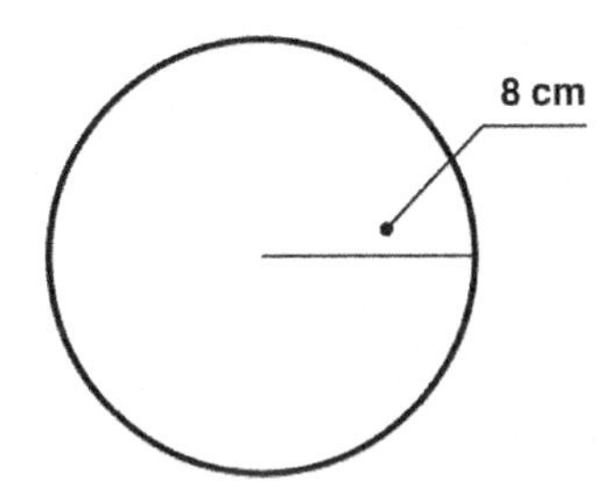

Data Analysis and Probability

Summarizing Data Presented Verbally, Tabularly, and Graphically

Tables, charts, and graphs are useful tools that convey information about different variables. They all organize, categorize, and compare data, and they come in different shapes and sizes. Each type has its own way of showing information, whether through a column, shape, or picture. To answer a question relating to a table, chart, or graph, some steps should be followed. First, read the problem thoroughly to determine what quantity is unknown. Then, read the title of the table, chart, or graph. The title should clarify what actual data the table is summarizing. Next, look at the key and both the horizontal and vertical axis labels, if they are given. These items will provide information about how the data is organized. Finally, look to see if there is any more labeling inside the table. Taking the time to get a good idea of what the table is summarizing will be helpful as it is used to interpret information.

Tables are a good way of showing a lot of information in a small space. The information in a table is organized in columns and rows. For example, a table may be used to show the number of votes each candidate received in an election. By interpreting the table, one may observe which candidate won the election and which candidates came in second and third.

The table below relates the number of items to the total cost. The table shows that one item costs $5. By looking at the table further, five items cost $25, ten items cost $50, and fifty items cost $250. This cost can be extended for any number of items. Since one item costs $5, then two items would cost $10. Though this information is not in the table, the given price can be used to calculate unknown information.

Number of Items	1	5	10	50
Cost ($)	5	25	50	250

A **bar graph** is a graph that summarizes data using bars of different heights. It is useful when comparing two or more items or when seeing how a quantity changes over time. It has both a horizontal and vertical axis. To interpret bar graphs, recognize what each bar represents and connect that to the two variables.

The bar graph below shows the scores for six people during three different games. The different colors of the bars distinguish between the three games, and the height of the bar indicates their score for that game. William scored 25 on game 3, and Abigail scored 38 on game 3. By comparing the bars, it is obvious that Williams scored lower than Abigail.

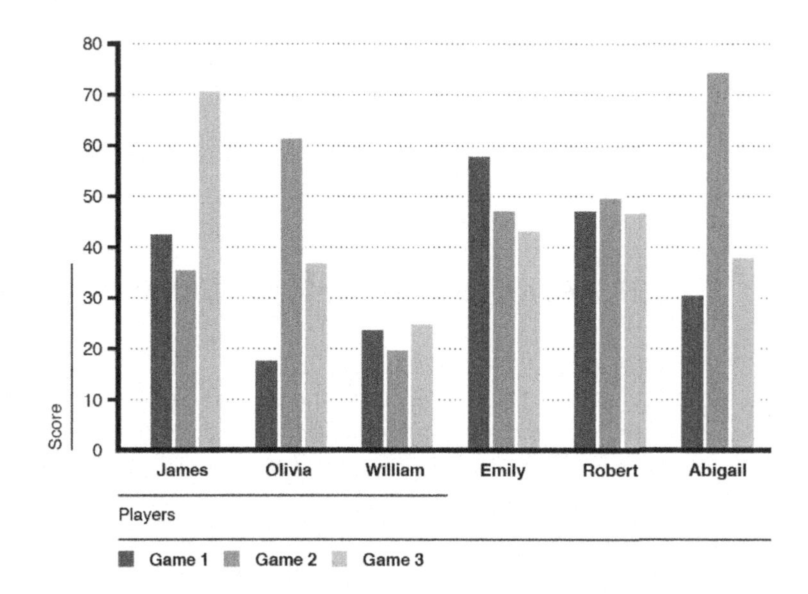

A **line graph** is a way to compare two variables that are plotted on opposite axes of a graph. The line indicates a continuous change as it rises or falls. The line's rate of change is known as its slope. The horizontal axis, or the *x*-axis, often represents a variable of time, and the vertical axis, or *y*-axis, represents another variable. If there are multiple lines, a comparison can be made between what the two lines represent. For example, the following line graph displays the change in temperature over five days. The top line represents the high, and the bottom line represents the low for each day. Looking at the top line alone, the high decreases on Tuesday, increases on Wednesday, decreases on Thursday, and increases again on Friday. The low temperatures have a similar trend, shown in the bottom line. The range in temperatures each day can be calculated by finding the difference between the top line and bottom line on a particular day. On Wednesday, the range was 14 degrees, from 62 to 76° F.

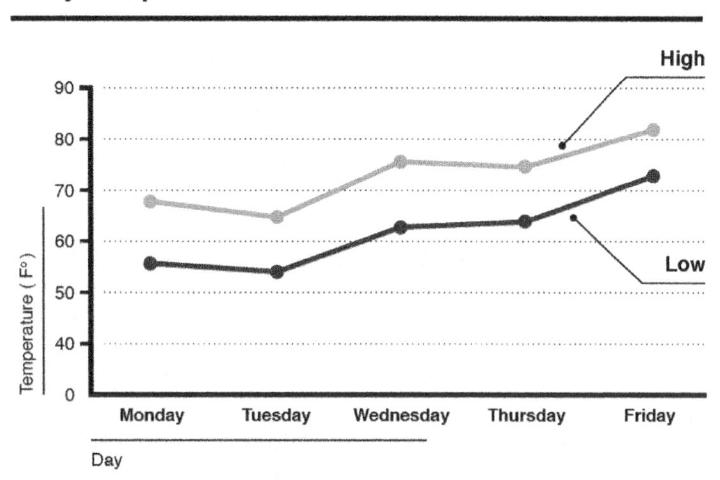

Pie charts show percentages of a whole; they are circular representations of data used to highlight numerical proportions. Each category represents a piece of the pie, and together, all of the pieces make up a whole. The size of each pie slice is proportional to the amount it represents; therefore, a reader can quickly make comparisons by visualizing the sizes of the pieces. The following pie chart is a simple example of three different categories shown in comparison to each other.

Light gray represents cats, dark gray represents dogs, and the medium shade of gray represents other pets. These three equal pieces each represent just more than 33 percent, or $\frac{1}{3}$ of the whole. In an example where the total pie represents 75,000 animals, then each category—dogs, cats, and other pets—would each be equal to $\frac{1}{3}$ of the total, or 25,000.

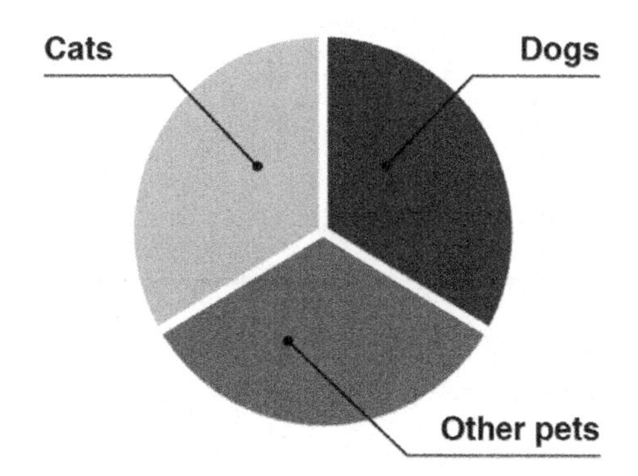

Stacked bar graphs are also used fairly frequently when comparing multiple variables at one time. They combine some elements of both pie charts and bar graphs, using the organization of bar graphs and the proportionality aspect of pie charts. The following is an example of a stacked bar graph that represents the number of students in a band playing drums, flute, trombone, and clarinet. Each bar graph is broken up further into girls and boys.

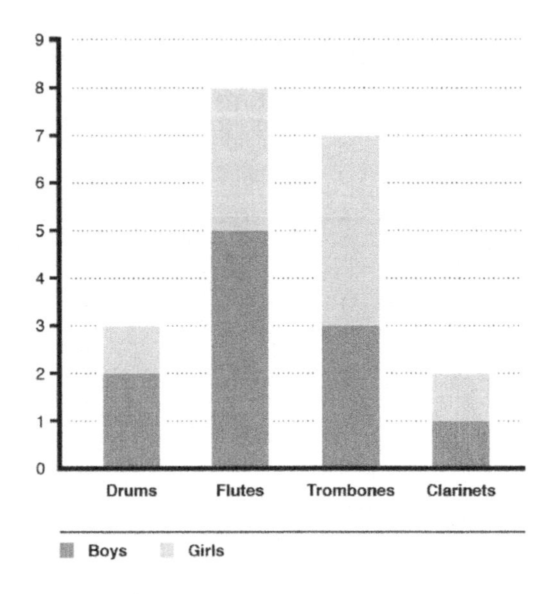

To determine how many boys play trombone, refer to the darker portion of the trombone bar, which indicates three boys.

A **scatterplot** is another way to represent paired data. It uses Cartesian coordinates, like a line graph, meaning it has both a horizontal and vertical axis. Each data point is represented as a dot on the graph. The dots are never connected with a line. For example, the following is a scatterplot showing the connection between people's ages and heights.

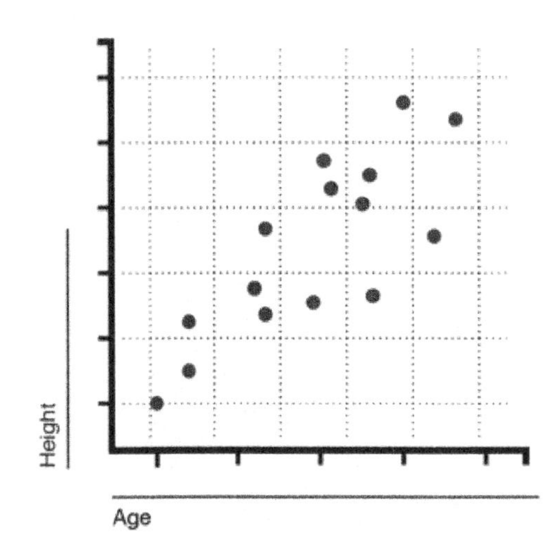

A scatterplot, also known as a **scattergram,** can be used to see if a correlation exists between a set of data. If the data resembles a straight line, then it is **associated,** or correlated. The following is an example of a scatterplot in which the data does not seem to have an association:

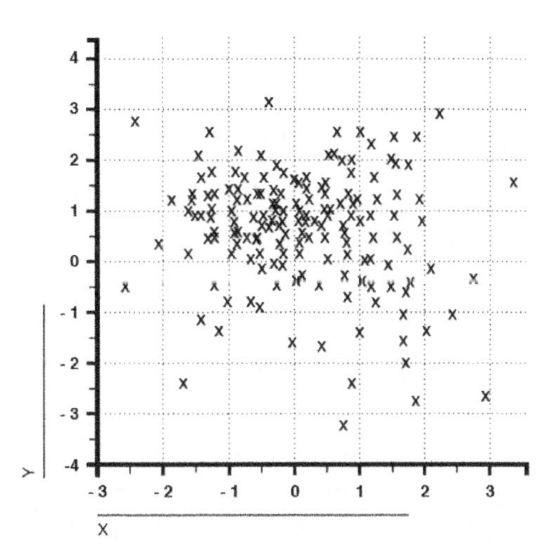

A **Venn diagram** represents each set of data as a circle. When the circles overlap, that means the sets of data overlap as well. A Venn diagram is also known as a *logic diagram* because it visualizes all possible logical combinations between two sets. Common elements of two sets are represented by the area of overlap. The following is an example of a Venn diagram of two sets A and B:

Parts of the Venn Diagram

A

Both
A and B

B

Another name for the area of overlap is the **intersection.** The intersection of A and B, written $A \cap B$, contains all elements that are in both sets A and B. The **union** of A and B, $A \cup B$, contains all elements in set A or set B (or both). Finally, the **complement** of $A \cup B$ is equal to all elements that are not in either set A or set B. These elements are placed outside of the circles.

The following is an example of a Venn diagram representing 24 students who were surveyed about their siblings. Ten students only had a brother, seven students only had a sister, and five had both a brother and a sister. Therefore, five is the intersection, represented by the section where the circles overlap. Two students had neither a brother nor a sister. Therefore, two is the complement and is placed outside of the circles.

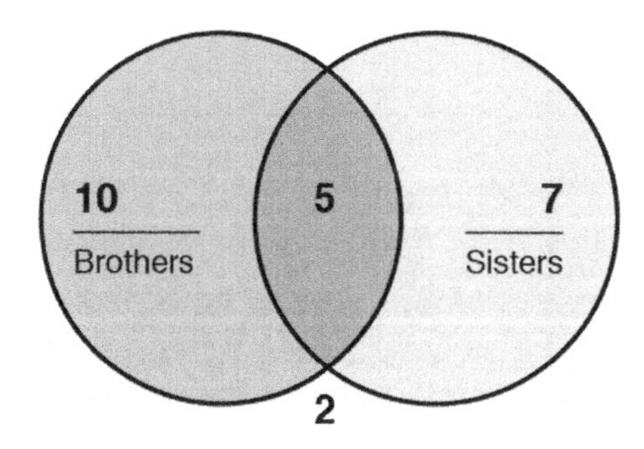

Venn diagrams can have more than two sets of data. The more circles, the more logical combinations that are represented by the overlapping. The following is a Venn diagram that represents sock colors worn by a class of students. There were 30 students surveyed. The innermost region represents students that had green, pink, and blue on their socks. Therefore, two students had all three colors. In this example, all students had at least one color on their socks, so there is no complement.

30 students

5
Green
5
3
Pink
2
3
2
10
Blue

Venn diagrams are typically not drawn to scale; however, if they are, and if each circle's area is proportional to the amount of data it represents, then it is called an **area-proportional Venn diagram**.

Recognizing Possible Associations and Trends in Data

The **independent variable** is the variable controlled by the experimenter. It stands alone and is not changed by other parts of the experiment. This variable is normally represented by x and is found on the horizontal, or x-axis, of a graph. The **dependent variable** changes in response to the independent variable. This variable is normally represented by y and is found on the vertical, or y-axis, of the graph. The relationship between two variables, x and y, can be seen on a scatterplot.

The following scatterplot shows the relationship between weight and height. The graph shows weight as x and height as y. The first dot on the left represents a person who is 45 kg and approximately 150 cm tall. The other dots correspond in the same way. As the dots move to the right and weight increases, height also increases. A line could be drawn through the middle of the dots to move from bottom left to top right. This line would indicate a **positive correlation** between the variables. If the variables had a **negative correlation**, then the dots would move from the top left to the bottom right.

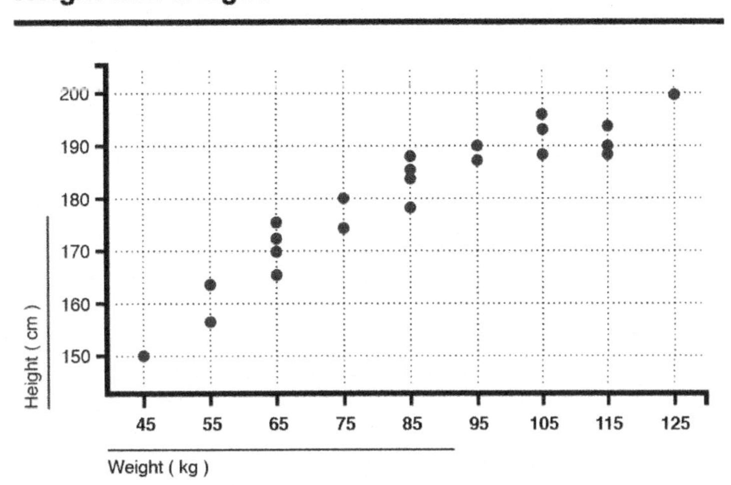

A scatterplot is useful in determining the relationship between two variables, but it is not required. Consider an example where a student scores a different grade on his math test for each week of the month. The independent variable would be the weeks of the month (time). The dependent variable would be the grades, because they change depending on the week. If the grades trended up as the weeks passed, then the relationship between grades and time would be positive. If the grades decreased as the time passed, then the relationship would be negative.

The relationship between two variables can further be described as strong or weak. The relationship between age and height shows a strong positive correlation because children grow taller as they grow up. In adulthood, the relationship between age and height becomes weak, and the dots will spread out. People stop growing in adulthood, and their final heights vary depending on factors like genetics and health. The closer the dots on the graph are to the trend line, the stronger the relationship. As they spread apart, the relationship becomes weaker. If they are too spread out to determine a trend (and thus, correlation), then the variables are said to have no correlation.

Identifying the Line of Best Fit

Data rarely fits into a straight line. Usually, we must be satisfied with approximations and predictions. Typically, when considering linear data with some sort of trend or association, the scatterplot for the data set appears to "fit" a straight line, and this line is known as the **line of best fit**. For instance, consider the following set of data which shows test scores for 10 students in a classroom on a final exam based on the number of hours each student studied: {(8,89), (7,88), (7,78), (7,77), (6,76), (5,76), (4,72), (5,65), (4,64), (3,61)}. Within each ordered pair, the first coordinate is the number of hours studied and the second coordinate is the test score. Graphing these in a scatterplot yields the following:

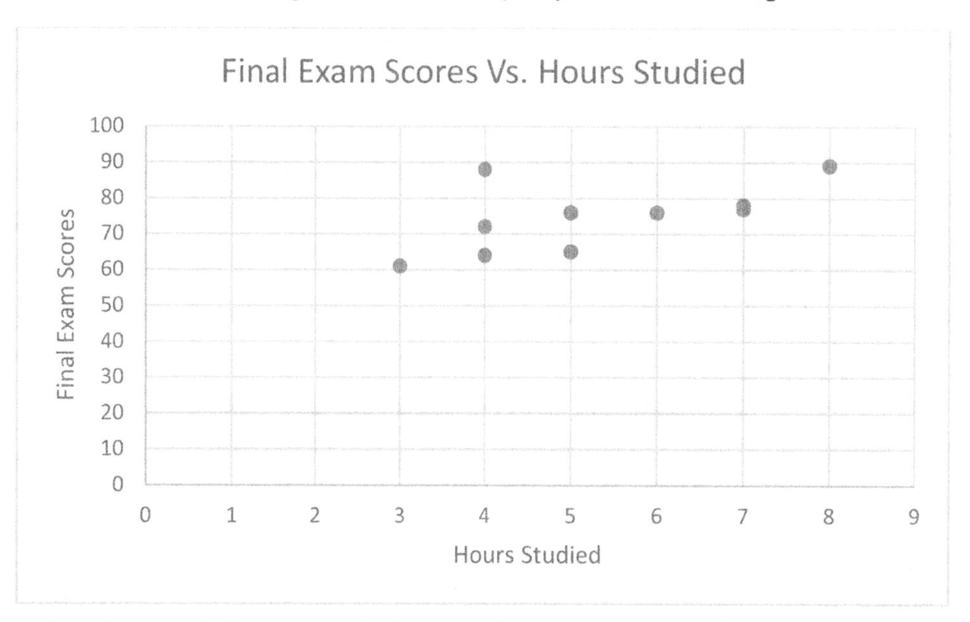

Note that the data does not follow a straight line exactly; however, a straight line can be drawn through the points as shown in the following plot:

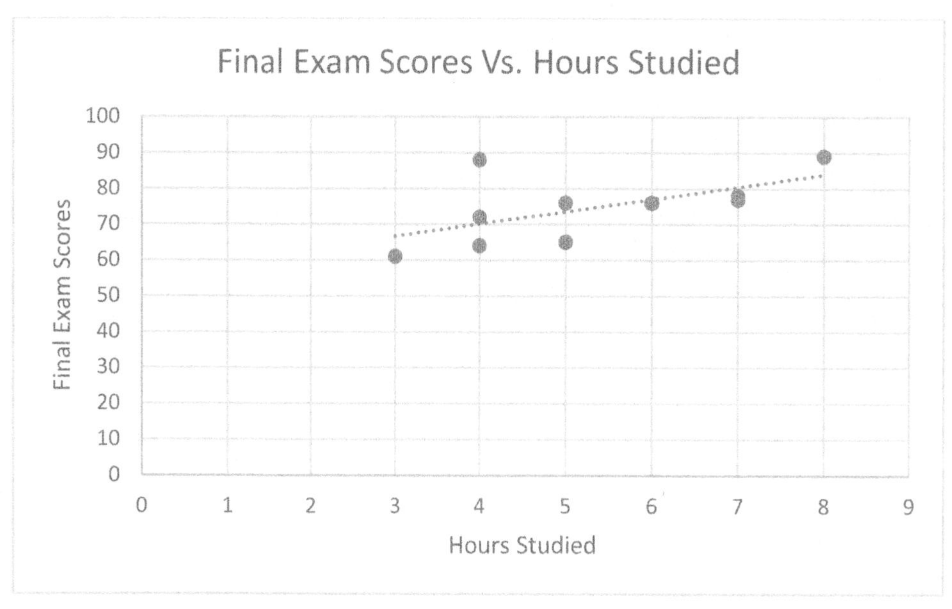

The inserted line is called the line of best fit, and it runs through the middle of the points. It can be found using technology, such as Excel® or a graphing calculator. In our example, Excel® was used for the scatterplot, and that function allowed for us to add a **linear trend line**, which is another name for the line of best fit. **Linear regression line** is another term that means line of best fit. Once you obtain the line, you can predict other points identifying other ordered pairs that the line runs through. For instance, the plot can be used to predict that a student who studied for 5.5 hours should receive close to a 75 on the final exam.

The distances between each point on the scatterplot and the line of best fit are called **residuals**. The line of best fit should attempt to minimize these distances. A line of best fit most accurately represents a set of data set if the residuals are small. The following graph shows a line of best fit that has very small residuals because the data points are all relatively close to the line:

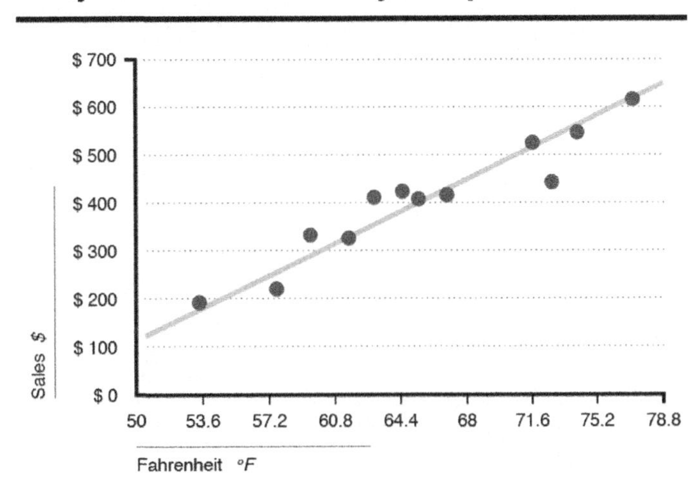

Here is one with larger residuals:

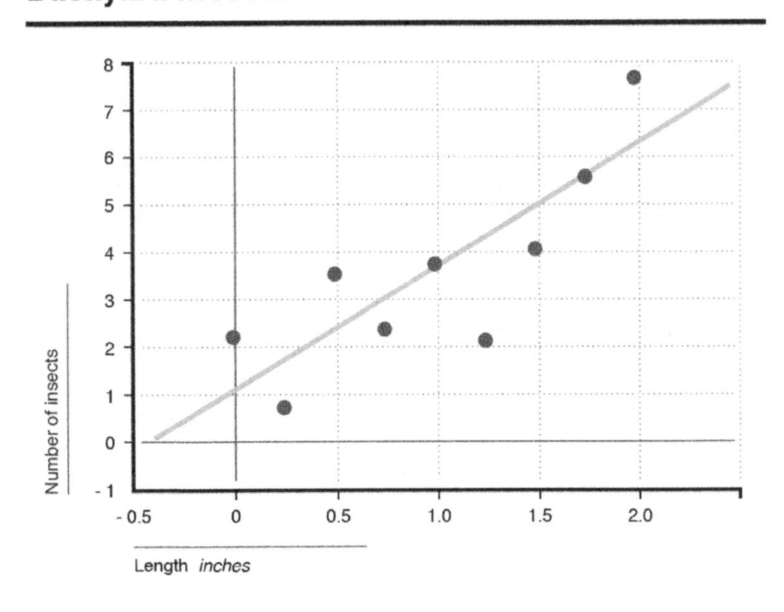

A data set must follow a linear pattern in order to have a line of best fit. Here is another example of a data set and the corresponding line of best fit:

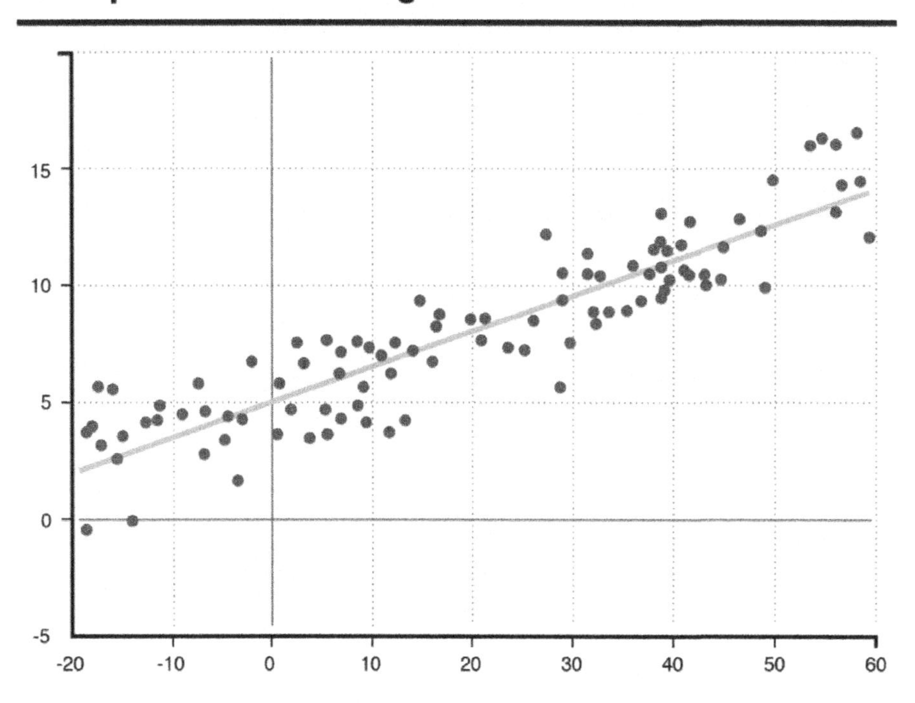

Finding the Probabilities of Single and Compound Events

Probability describes how likely it is that an event will occur. Probabilities are always a number from 0 to 1. If an event has a high likelihood of occurrence, it will have a probability close to 1. If there is only a small chance that an event will occur, the likelihood is close to 0. A fair six-sided die has one of the numbers 1, 2, 3, 4, 5, and 6 on each side. When this die is rolled, there is a one in six chance that it will land on 2. This is because there are six possibilities and only one side has a 2 on it. The probability then is $\frac{1}{6}$ or .167. The probability of rolling an even number from this die is three in six, or $\frac{1}{2}$, or .5. This is because there are three sides on the die with even numbers (2, 4, 6), and there are six possible sides. The probability of rolling a number less than 10 is 1; since every side of the die has a number less than 10, it would be impossible to roll a number 10 or higher. On the other hand, the probability of rolling a number larger than 20 is zero since there are no numbers greater than 20 on the die.

If a teacher says that the probability of anyone passing her final exam is .2, is it highly likely that anyone will pass? No, the probability of anyone passing her exam is low because .2 is closer to 0 than to 1. If another teacher is proud that the probability of students passing his class is .95, how likely is it that a student will pass? It is highly likely that a student will pass because the probability, .95, is very close to 1.

A **probability experiment** is a repeated action that has a specific set of possible results. The result of such an experiment is known as an **outcome**, and the set of all potential outcomes is known as the **sample space.** An **event** consists of one or more of those outcomes. For example, consider the probability experiment of tossing a coin and rolling a six-sided die. The coin has two possible outcomes—a heads or a tails—and the die has six possible outcomes—rolling each number 1–6. Therefore, the sample space has twelve possible outcomes: a heads or a tails paired with each roll of the die.

A **simple event** is an event that consists of a single outcome. For instance, selecting a queen of hearts from a standard fifty-two-card deck is a simple event; however, selecting a queen is not a simple event because there are four possibilities.

Classical, or **theoretical, probability** is when each outcome in a sample space has the same chance to occur. The probability for an event is equal to the number of outcomes in that event divided by the total number of outcomes in the sample space. For example, consider rolling a six-sided die. The probability of rolling a 2 is $\frac{1}{6}$, and the probability of rolling an even number is $\frac{3}{6}$, or $\frac{1}{2}$, because there are three even numbers on the die. This type of probability is based on what should happen in theory but not what actually happens in real life.

Empirical probability is based on actual experiments or observations. For example, if a die is rolled eight times, and a 1 is rolled two times, the empirical probability of rolling a 1 is $\frac{2}{8} = \frac{1}{4}$, which is higher than the theoretical probability. The Law of Large Numbers states that as an experiment is completed repeatedly, the empirical probability of an event should get closer to the theoretical probability of an event.

The **addition rule** is necessary to find the probability of event *A* or event *B* occurring *or* both occurring at the same time. If events *A* and *B* are **mutually exclusive** or **disjoint,** which means they cannot occur at the same time, then:

$$P(A \text{ or } B) = P(A) + P(B)$$

If events *A* and *B* are not mutually exclusive,

$$P(A \text{ or } B) = P(A) + P(B) - P(A \text{ and } B)$$

Where $P(A \text{ and } B)$ represents the probability of event *A* and *B* both occurring at the same time. An example of two events that are mutually exclusive are rolling a 6 on a die and rolling an odd number on a die. The probability of rolling a 6 or rolling an odd number is:

$$\frac{1}{6} + \frac{3}{6} = \frac{4}{6} = \frac{2}{3}$$

Rolling a 6 and rolling an even number are not mutually exclusive because there is some overlap. The probability of rolling a 6 or rolling an even number is:

$$\frac{1}{6} + \frac{3}{6} - \frac{1}{6} = \frac{3}{6} = \frac{1}{2}$$

Conditional Probability

The **multiplication rule** is necessary when finding the probability that event *A* occurs in a first trial and event *B* occurs in a second trial, which is written as $P(A \text{ and } B)$. This rule differs if the events are independent or dependent. Two events *A* and *B* are **independent** if the occurrence of one event does not affect the probability that the other will occur. If *A* and *B* are not independent, they are **dependent,** and the outcome of the first event somehow affects the outcome of the second. If events A and B are independent,

$$P(A \text{ and } B) = P(A)P(B)$$

If events A and B are dependent,

$$P(A \text{ and } B) = P(A)P(B|A)$$

Where $P(B|A)$ represents the probability event B occurs given that event A has already occurred.

$P(B|A)$ represents **conditional probability**, or the probability of event B occurring given that event A has already occurred. $P(B|A)$ can be found by dividing the probability of events A and B both occurring by the probability of event A occurring using the formula:

$$P(B|A) = \frac{P\ (A \text{ and } B)}{P(A)}$$

This represents the total number of outcomes remaining for B to occur after A occurs. This formula is derived from the multiplication rule with dependent events by dividing both sides by $P(A)$. Note that $P(B|A)$ and $P(A|B)$ are not the same. The first quantity shows that event B has occurred after event A, and the second quantity shows that event A has occurred after event B. Incorrectly interchanging these ideas is known as **confusing the inverse.**

Consider the case of drawing two cards from a deck of fifty-two cards. The probability of pulling two queens would vary based on whether the initial card was placed back in the deck for the second pull. If the card is placed back in, the probability of pulling two queens is:

$$\frac{4}{52} \times \frac{4}{52} = 0.00592$$

If the card is not placed back in, the probability of pulling two queens is:

$$\frac{4}{52} \times \frac{3}{51} = 0.00452$$

When the card is not placed back in, both the numerator and denominator of the second probability decrease by 1. This is due to the fact that there is one less queen in the deck, and there is one less total card in the deck as well.

Conditional probability is used frequently when probabilities are calculated from tables. A **two-way frequency table** displays categorical data with two variables, and it highlights relationships that exist between those two variables. Such tables are used frequently to summarize survey results and are also known as **contingency tables**. Each cell shows a count pertaining to that individual variable pairing, known as a **joint frequency**, and the totals of each row and column also are in the table.

Consider the following two-way frequency table:

Distribution of the Residents of a Particular Village

	70 or older	69 or younger	Totals
Women	20	40	60
Men	5	35	40
Total	25	75	100

The table shows the breakdown of ages and sexes of 100 people in a particular village. The end of each row or column displays the number of people represented by the corresponding data, and the total number of people is shown in the bottom right corner. For instance, there were 25 people aged 70 or older and 60 women in the data. The 20 in the first cell shows that out of 100 total villagers, 20 were women aged 70 or older. The 5 in the cell below shows that out of 100 total villagers, 5 were men aged 70 or older.

A two-way table can also show relative frequencies by indicating the percentages of people instead of the count. If each frequency is calculated over the entire total of 100, the first cell would be 20% or 0.2. However, the relative frequencies can also be calculated over row or column totals. If row totals were used, the first cell would be:

$$\frac{20}{60} = 0.333 = 33.3\%$$

If column totals were used, the first cell would be:

$$\frac{20}{25} = 0.8 = 80\%$$

Such tables can be used to calculate conditional probabilities. Consider a randomly selected villager. The probability of selecting a male 70 years old or older is $\frac{5}{100} = 0.05$ because there are 5 males over the age of 70 and 100 total villagers.

Approximating the Probability of a Chance Event

Probability problems require knowledge of the total number of events in the sample space. Different methods can be used to count the number of possible outcomes, depending on whether different arrangements of the same items are counted only once or separately. **Permutations** are arrangements in which different sequences are counted separately. Therefore, order matters in permutations. **Combinations** are arrangements in which different sequences are not counted separately. Therefore, order does not matter in combinations. Permutations would consider 123 different from 321; combinations, on the other hand, consider 123 the same as 321.

If the sample space contains *n* different permutations of *n* different items and all of them must be selected, there are $n!$ different possibilities, with $n!$ Referring to an integer *n* multiplied by every positive integer less than it. For example, five different books can be rearranged 5! = 120 times (5 x 4 x 3 x 2 x 1). The probability of one person randomly ordering those five books in the same way as another person is $\frac{1}{120}$. A different calculation is necessary if a number less than *n* is to be selected or if order does not matter. In general, the notation $P(n, r)$ represents the number of ways to arrange *r* objects from a set of *n* if order does matter, and:

$$P(n,r) = \frac{n!}{(n-r)!}$$

Therefore, in order to calculate the number of ways five books can be arranged in three slots if order matters, plug *n* = 5 and *r* = 3 in the formula to obtain:

$$P(5,3) = \frac{5!}{(5-3)!} = \frac{5!}{2!} = 60$$

Secondly, $C(n, r)$ represents the total number of *r* combinations selected out of *n* items when order does not matter, and:

$$C(n,r) = \frac{n!}{(n-r)!\ r!}$$

Therefore, the number of ways five books can be arranged in three slots if order does not matter is:

$$C(5,3) = \frac{5!}{(5-3)!\ 3!} = 10$$

The following relationship exists between permutations and combinations:

$$C(n,r) = \frac{P(n,r)}{r!}$$

Developing and Using a Probability Model

A **discrete random variable** is a set of values that is either finite or countably infinite. If there are infinitely many values, being *countable* means that each individual value can be paired with a natural number. For example, the number of coin tosses before getting heads could potentially be infinite, but the total number of tosses is countable. Each toss refers to a number, like the first toss, second toss, etc. A **continuous random variable** has infinitely many values that are not countable. The individual items cannot be enumerated; an example of such a set is any type of measurement. There are infinitely many heights of human beings due to decimals that exist within each inch, centimeter, millimeter, etc. Each type of variable has its own **probability distribution**, which calculates the probability for each potential value of the random variable. Probability distributions exist in tables, formulas, or graphs. The expected value of a random variable represents what the mean value should be in either a large sample size or after many trials. According to the Law of Large Numbers, after many trials, the actual mean and that of the probability distribution should be very close to the expected value. The **expected value** is a weighted average that is calculated as $E(X) = \sum x_i p_i$, where x_i represents the value of each outcome, and p_i represents the probability of each outcome.

The expected value if all of the probabilities are equal is:

$$E(X) = \frac{x_1 + x_2 + \cdots + x_n}{n}$$

Expected value is often called the **mean of the random variable** and is known as a **measure of central tendency** like mean and mode.

A **binomial probability distribution** adheres to some important criteria: it must consist of a fixed number of trials where all trials are independent, each trial must have an outcome classified as either success or failure, and the probability of a success must be the same in each trial. Within any binomial experiment, *x* is the number of resulting successes, *n* is the number of trials, *P* is the probability of success within each trial, and $Q = 1 - P$ is the probability of failure within each trial. The probability of obtaining *x* successes within *n* trials is:

$$\binom{n}{x} P^x (1 - P)^{n-x}$$

The expression below is called the **binomial coefficient**.:

$$\binom{n}{x} = \frac{n!}{x!\,(n - x)!}$$

A binomial probability distribution could be used to find the probability of obtaining exactly two heads on five tosses of a coin. In the formula, *x* = 2, *n* = 5, *P* = 0.5, and *Q* = 0.5.

A **uniform probability distribution** exists when there is constant probability. Each random variable has equal probability, and its graph looks a rectangle because the height, representing the probability, is constant.

Finally, a **normal probability distribution** has a graph that is symmetric and bell-shaped; an example using body weight is shown here:

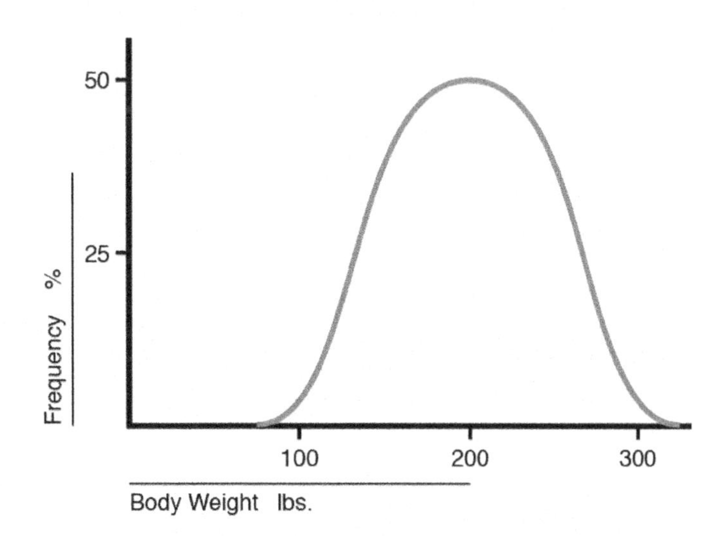

Population percentages can be estimated using normal distributions. For example, the probability that a data point will be less than the mean is 50 percent. The Empirical Rule states that 68 percent of the data falls within 1 standard deviation of the mean, 95 percent falls within 2 standard deviations of the mean, and 99.7 percent falls within 3 standard deviations of the mean. A **standard normal distribution** is a normal distribution with a mean equal to 0 and standard deviation equal to 1. The area under the entire curve of a standard normal distribution is equal to 1.

Using Measures of Central Tendency to Draw Inferences About Populations

The three most common calculations for a set of data are the mean, median, and mode. These three are called **measures of central tendency**, which are helpful in comparing two or more different sets of data. The **mean** refers to the average and is found by adding up all values and dividing the total by the number of values. In other words, the mean is equal to the sum of all values divided by the number of data entries. For example, if you bowled a total of 532 points in 4 bowling games, your mean score was $\frac{532}{4} = 133$ points per game. Students can apply the concept of mean to calculate what score they need on a final exam to earn a desired grade in a class.

The **median** is found by lining up values from least to greatest and choosing the middle value. If there is an even number of values, then calculate the mean of the two middle amounts to find the median. For example, the median of the set of dollar amounts $5, $6, $9, $12, and $13 is $9. The median of the set of dollar amounts $1, $5, $6, $8, $9, $10 is $7, which is the mean of $6 and $8.

The **mode** is the value that occurs the most. The mode of the data set {1, 3, 1, 5, 5, 8, 10} actually refers to two numbers: 1 and 5. In this case, the data set is **bimodal** because it has two modes. A data set can have no mode if no amount is repeated.

Another useful statistic is range. The **range** for a set of data refers to the difference between the highest and lowest value.

In some cases, numbers in a list of data might have weights attached to them. In that case, a **weighted mean** can be calculated. A common application of a weighted mean is GPA. In a semester, each class is assigned a number of credit hours (its weight), and at the end of the semester each student receives a grade. To compute GPA, an A is a 4, a B is a 3, a C is a 2, a D is a 1, and an F is a 0. Consider a student that takes a 4-hour English class, a 3-hour math class, and a 4-hour history class and receives all B's. The weighted mean, GPA, is found by multiplying each grade times its weight, number of credit hours, and dividing by the total number of credit hours. Therefore, the student's GPA is:

$$\frac{(3 \times 4) + (3 \times 3) + (3 \times 4)}{11} = \frac{33}{11} = 3.0$$

The following bar chart shows how many students attend a cycle class on each day of the week. To find the mean attendance for the week, add each day's attendance together:

$$10 + 7 + 6 + 9 + 8 + 14 + 4 = 58$$

Then divide the total by the number of days:

$$58 \div 7 = 8.3$$

The mean attendance for the week was 8.3 people. The median attendance can be found by putting the attendance numbers in order from least to greatest: 4, 6, 7, 8, 9, 10, 14, and choosing the middle number: 8 people. This set of data has no mode because no numbers repeat. The range is 10, which is found by finding the difference between the lowest number, 4, and the highest number, 14.

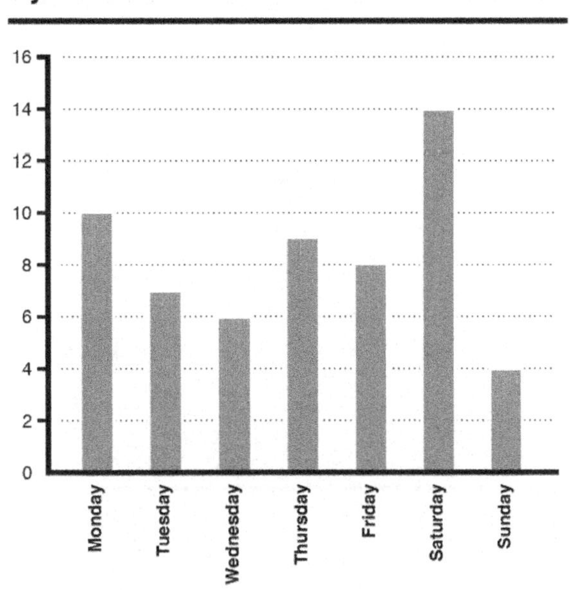

A **histogram** is a bar graph used to group data into "bins" that cover a range on the x-axis. Histograms consist of rectangles whose heights are equal to the frequency of a specific category. The horizontal axis represents the specific categories. Because they cover a range of data, these bins have no gaps between bars, unlike the bar graph above. In a histogram showing the heights of adult golden retrievers, the bottom axis would be groups of heights, and the y-axis would be the number of dogs in each range. Evaluating this histogram would show the height of most golden retrievers as falling within a certain range. It also provides information to find the average height and range for how tall golden retrievers may grow.

On the following histogram, the horizontal axis represents ranges of the points scored on an exam, and the vertical axis represents the number of students who scored that grade. For example, approximately 33 students scored in the 60 to 70 range.

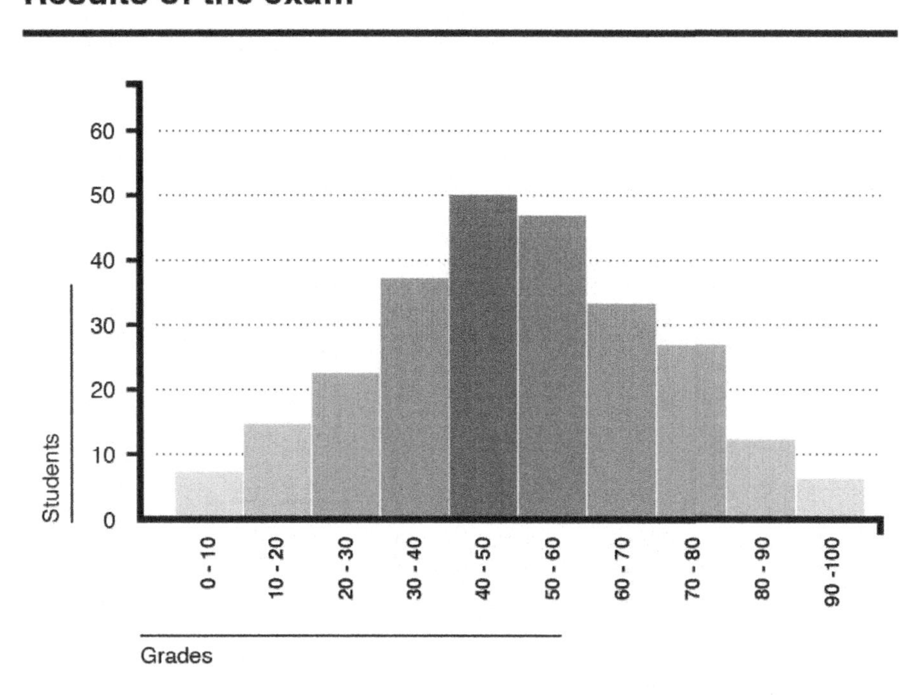

Histograms can be classified as having data **skewed to the left, skewed to the right**, or **normally distributed**, which is also known as **bell-shaped**. These three classifications can be seen in the following chart:

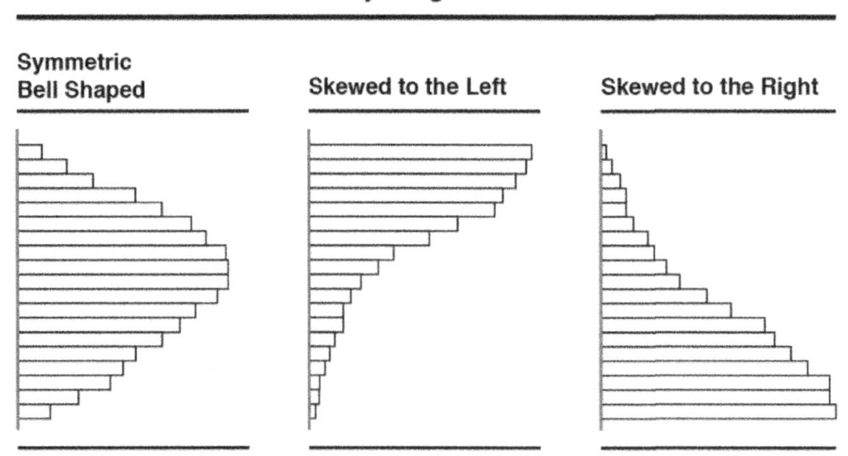

When the data is normal, the mean, median, and mode are very similar because they all represent the most typical value in the data set. In this case, the mean is typically considered the best measure of central tendency because it includes all data points. However, if the data is skewed, the mean becomes less meaningful because it is dragged in the direction of the skew. Therefore, the median becomes the best measure because it is not affected by any outliers.

The measures of central tendency and the range may also be found by evaluating information on a line graph.

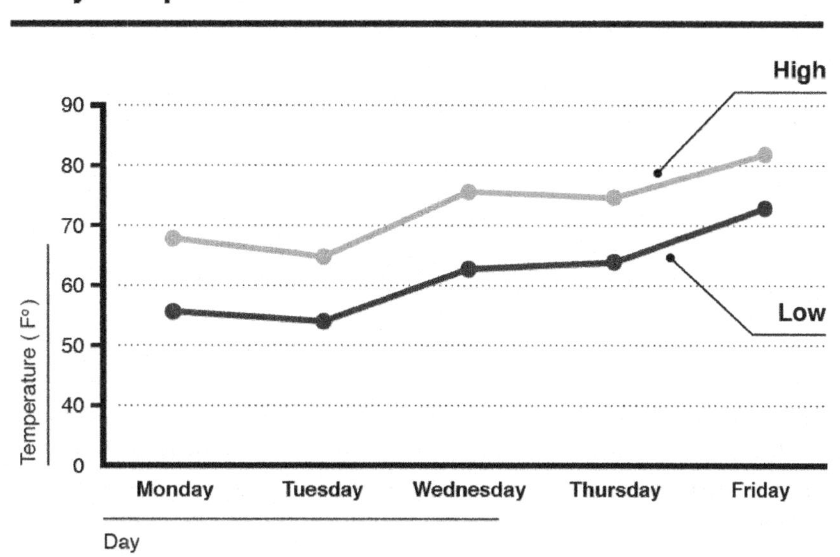

In the line graph above that shows the daily high and low temperatures, the average high temperature can be found by gathering data from each day. The days' highs are 69, 65, 75, 74, and 81. To find the average, add them together to get 364, then divide by 5 (because there are 5 temperatures). The average high for the five days is 72.8. If 72.8 degrees is found on the graph, it will fall in the middle of all the values. The low temperature can be found in the same way.

Determining How Changes in Data Affect Measures of Central Tendency

An **outlier** is a data point that lies an unusual distance from other points in the data set. Removing an outlier from a data set will change the measures of central tendency. Removing a *large outlier* (a high number) from a data set will decrease both the mean and the median. Removing a *small outlier* (a number much lower than most in the data set) from a data set will increase both the mean and the median.

For example, given the data set {3, 6, 8, 12, 13, 14, 60}, the data point 60 is an outlier because it is unusually far from the other points. In this data set, the mean is 16.6. Notice that this mean number is even larger than all other data points in the set except for 60. Removing the outlier changes the mean to 9.3, and the median goes from 12 to 10. Removing an outlier will also decrease the range. In the data set above, the range is 57 when the outlier is included, but it decreases to 11 when the outlier is removed.

What would happen with an extremely low value in a data set like this one: {12, 87, 90, 95, 98, 100}? The mean of the given set is 80. When the outlier, 12, is removed, the mean should increase and fit more closely to the other data points. Removing 12 and recalculating the mean show that this is correct. After removing the outlier, the mean is 94. So, removing a large outlier will decrease the mean while removing a small outlier will increase the mean.

Adding an outlier to a data set will also affect the measures of central tendency. When a larger outlier is added to a data set, the mean and median increase. When a small outlier is added to a data set, the mean and median decrease. Adding an outlier to a data set will increase the range.

Using Statistics to Gain Information About a Population

Statistics is the branch of mathematics that deals with the collection, organization, and analysis of data. A statistical question is one that can be answered by collecting and analyzing data. When collecting data, expect variability. For example, "How many pets does Yanni own?" is not a statistical question because it can be answered in one way. "How many pets do the people in a certain neighborhood own?" is a statistical question because, to determine this answer, one would need to collect data from each person in the neighborhood, and it is reasonable to expect the answers to vary.

Identify the following questions as statistical or not statistical:

- How old are you?
- What is the average age of the people in your class?
- How tall are the students in Mrs. Jones' sixth grade class?
- Do you like Brussels sprouts?

The first and last questions are not statistical, but the two middle questions are.

Data collection can be done through surveys, experiments, observations, and interviews. A **census** is a type of survey that is done with a whole population. Because it can be difficult to collect data for an entire population, sometimes a **sample** is used. In this case, one would survey only a fraction of the population and make inferences about the data. Sample surveys are not as accurate as a census, but they are an easier and less expensive method of collecting data. An **experiment** is used when a researcher wants to explain how one variable causes change in another variable. For example, if a researcher wanted to know if a particular drug affects weight loss, he or she would choose a **treatment group** that would take the drug, and another group, the **control group**, that would not take the drug. Special care must be taken when choosing these groups to ensure that bias is not a factor. **Bias** occurs when an outside factor influences the outcome of the research. In observational studies, the researcher does not try to influence either variable but simply observes the behavior of the subjects. Interviews are sometimes used to collect data as well. The researcher will ask questions that focus on his or her area of interest in order to gain insight from the participants. When gathering data through observation or interviews, it is important that the researcher is well trained so that he or she does not influence the results and the study remains reliable. A study is reliable if it can be repeated under the same conditions and the same results are received each time.

Using Random Sampling to Draw Inferences About a Population

In statistics, a **population** contains all subjects being studied. For example, a population could be every student at a university or all males in the United States. A **sample** consists of a group of subjects from an entire population. A sample would be 100 students at a university or 100,000 males in the United States. **Inferential statistics** is the process of using a sample to generalize information concerning populations. **Hypothesis testing** is the actual process used when evaluating claims made about a population based on a sample.

A **statistic** is a measure obtained from a sample, and a **parameter** is a measure obtained from a population. For example, the mean SAT score of the 100 students at a university would be a statistic, and the mean SAT score of all university students would be a parameter.

The beginning stages of hypothesis testing starts with formulating a **hypothesis,** a statement made concerning a population parameter. The hypothesis may be true, or it may not be true. The experiment will help answer that question. In each setting, there are two different types of hypotheses: the **null hypothesis**, written as H_0, and the **alternative hypothesis**, written as H_1. The null hypothesis represents verbally when there is not a difference between two parameters, and the alternative hypothesis represents verbally when there is a difference between two parameters. Consider the following experiment: A researcher wants to see if a new brand of allergy medication has any effect on drowsiness of the patients who take the medication. He wants to know if the average hours spent sleeping per day increases. The mean for the population under study is 8 hours, so $\mu = 8$. In other words, the population parameter is μ, the mean. The null hypothesis is $\mu = 8$ and the alternative hypothesis is $\mu > 8$. When using a smaller sample of a population, the null hypothesis represents the situation when the mean remains unaffected and the alternative hypothesis represents the situation when the mean increases. The chosen statistical test will apply the data from the sample to actually decide whether the null hypothesis should or should not be rejected.

Practice Test

Spelling

For each of the following questions, choose the one answer that is misspelled.

1.

a. Cacofony
b. Beige
c. Apathy
d. Courage

2.

a. Temporary
b. Antham
c. Statute
d. Reasonable

3.

a. Unpleasant
b. Jaring
c. Corporate
d. Roulette

4.

a. Marred
b. Cajole
c. Blanketed
d. Sinphony

5.

a. Statement
b. Balance
c. Minimum
d. Transfered

6.

a. Blasphamy
b. Heretic
c. Religious
d. Canon

7.

a. Courage
b. Hopeful
c. Freightened
d. Mistaken

8.

a. Difficult
b. Intense
c. Ascetic
d. Managable

9.

a. Urgent
b. Combersome
c. Deficit
d. Language

10.

a. Minute
b. Caripace
c. Stringent
d. Sedition

11.

a. Hipocracy
b. Legend
c. Dominion
d. History

12.

a. Milieiu
b. Riptide
c. Sacrament
d. Blatant

13.

a. Carriage
b. Decorative
c. Ambiant
d. Misshapen

14.

a. Dangerous
b. Decedent
c. Plagiarism
d. Despondant

15.

a. Plaintive
b. Rambuncteous
c. Banal
d. Finale

16.
a. Despotte
b. Longitude
c. Climate
d. Category

17.
a. Laborious
b. Cateclysmic
c. Tradesman
d. Transition

18.
a. Dissolution
b. Contractual
c. Abhorent
d. Seismic

19.
a. Rudamentary
b. Advantageous
c. Torpor
d. Tumultuous

20.
a. Succession
b. Empediment
c. Seizure
d. Electricity

21.
a. Sacrosanct
b. Judicious
c. Topple
d. Antithisis

22.
a. Judiscious
b. Eliminate
c. Bourgeoisie
d. Exclamatory

23.
a. Incredulous
b. Annoint
c. Tableau
d. Missile

24.

a. Ableist
b. Dexterous
c. Incognito
d. Misscommunication

25.

a. Archery
b. Mathmetician
c. Incongruous
d. Congressional

26.

a. Magnanimous
b. Turbulent
c. Delicatesin
d. Ardor

27.

a. Detritous
b. Siphon
c. Category
d. Relegate

28.

a. Riboflavin
b. Mangy
c. Superlative
d. Delerium

29.

a. Podcast
b. Dapper
c. Spontanaity
d. Arduous

30.

a. Malecotent
b. Execute
c. Southerly
d. Catholic

31.

a. Ambient
b. Localized
c. Ammunition
d. Vaxinated

32.

a. Purplexed
b. Superfluous
c. Candidate
d. Incandescent

33.

a. Collapse
b. Eccological
b. Realization
d. Mundane

34.

a. Economist
b. Publication
c. Negociable
d. Cartridge

35.

a. Piutocrat
b. Cathartic
c. Greenery
d. Hibiscus

36.

a. Demonstretive
b. Artisanal
c. Umbilical
d. Collaborative

37.

a. Biological
b. Tertiery
c. Robust
d. Endemic

38.

a. Systemic
b. Irrigation
c. Furtive
d. Etamology

39.

a. Comprehensive
b. Optimism
c. Idealogical
d. Iterative

40.

a. Emmersion
b. Punitive
c. Explanation
d. Transcription

41.

a. Succumb
b. Nefareous
c. Renewable
d. Credible

42.

a. Incontravertible
b. Discussion
c. Impurity
d. Disengagement

43.

a. Academia
b. Workshop
c. Approppo
d. Opportunities

44.

a. Machination
b. Revival
c. Adversariel
d. Acquisition

45.

a. Development
b. Acceptable
c. Training
d. Lagastician

46.

a. Connectivity
b. Telocommute
c. Visualization
d. Consecrate

47.

a. Clairivoyant
b. Clearance
c. Xylophone
d. Zoonotic

48.

a. Technocratic
b. Retroffit
c. Corporate
d. Hexagon

49.

a. Leaflet
b. Evident
c. Mercurial
d. Hydroplain

50.

a. Councelor
b. Reservation
c. Buoy
d. Gargantuan

Vocabulary

For each of the following questions, choose the one word whose meaning is MOST similar to the word printed in capital letters.

1. WARY

a. Religious
b. Adventurous
c. Tired
d. Negligent
e. Cautious

2. PROXIMITY

a. Estimate
b. Delicate
c. Precarious
d. Splendor
e. Closeness

3. PARSIMONIOUS

a. Lavish
b. Harmonious
c. Miserly
d. Careless
e. Generous

4. PROPRIETY

a. Ownership
b. Appropriateness
c. Patented
d. Abstinence
e. Sobriety

5. BOON
    a. Cacophony
    b. Hopeful
    c. Benefit
    d. Squall
    e. Omen

6. STRIFE
    a. Plague
    b. Industrial
    c. Picketinq
    d. Eliminate
    e. Conflict

7. QUALM
    a. Calm
    b. Uneasiness
    c. Assertion
    d. Pacify
    e. Victory

8. VALOR
    a. Rare
    b. Coveted
    c. Leadership
    d. Bravery
    e. Honorable

9. ZEAL
    a. Craziness
    b. Resistance
    c. Fervor
    d. Opposition
    e. Apprehension

10. EXTOL
    a. Glorify
    b. Demonize
    c. Chide
    d. Admonish
    e. Criticize

11. REPROACH
    a. Locate
    b. Blame
    c. Concede
    d. Orate
    e. Honor

12. MILIEU
    a. Bacterial
    b. Damp
    c. Ancient
    d. Environment
    e. Uncertain

13. GUILE
    a. Masculine
    b. Stubborn
    c. Naïve
    d. Gullible
    e. Deception

14. ASSENT
    a. Acquiesce
    b. Climb
    c. Assert
    d. Demand
    e. Heighten

15. DEARTH
    a. Grounded
    b. Scarcity
    c. Lethal
    d. Risky
    e. Hearty

16. CONSPICUOUS
    a. Scheme
    b. Obvious
    c. Secretive
    d. Ballistic
    e. Paranoid

17. ONEROUS
    a. Responsible
    b. Generous
    c. Hateful
    d. Burdensome
    e. Wealthy

18. BANAL
    a. Inane
    b. Novel
    c. Painful
    d. Complimentary
    e. Inspired

19. CONTRITE
    a. Tidy
    b. Unrealistic
    c. Contrived
    d. Corrupt
    e. Remorseful

20. MOLLIFY
    a. Pacify
    b. Blend
    c. Negate
    d. Amass
    e. Emote

21. CAPRICIOUS
    a. Skillful
    b. Sanguine
    c. Chaotic
    d. Fickle
    e. Agreeable

22. PALTRY
    a. Appealing
    b. Worthy
    c. Trivial
    d. Fancy
    e. Disgusting

23. SHIRK
    a. Counsel
    b. Evade
    c. Diminish
    d. Sharp
    e. Annoy

24. ASSUAGE
    a. Irritate
    b. Persuade
    c. Argue
    d. Redirect
    e. Soothe

25. TACIT
    a. Unspoken
    b. Shortened
    c. Tenuous
    d. Regal
    e. Timid

26. ADVOCATE
   a. Entice
   b. Brandish
   c. Decline
   d. Support
   e. Align

27. CORROBORATE
   a. Verify
   b. Deny
   c. Forget
   d. Claim
   e. Create

28. SHEER
   a. Opaque
   b. Muddy
   c. Brave
   d. Depressed
   e. Translucent

29. VOW
   a. Liberate
   b. Worship
   c. Lie
   d. Break
   e. Promise

30. INCITE
   a. Calm
   b. Provoke
   c. Smell
   d. Repent
   e. Weaken

Analogies

For each of the following questions, choose the option that best completes the sentence.

1. **Chapter** is to **book** as
   a. Book is to story.
   b. Fable is to myth.
   c. Paragraph is to essay.
   d. Dialogue is to play.
   e. Story is to tale.

2. **Dress** is to **garment** as
   a. Diesel is to fuel.
   b. Month is to year.
   c. Suit is to tie.
   d. Clothing is to wardrobe.
   e. Coat is to winter.

3. **Car** is to **garage** as **plane** is to
   a. Sky
   b. Passenger.
   c. Airport.
   d. Runway.
   e. Hanger.

4. **Trickle** is to **gush** as
   a. Bleed is to cut.
   b. Rain is to snow.
   c. Tepid is to scorching.
   d. Sob is to sniffle.
   e. Ocean is to river.

5. **Acne** is to **dermatologist** as **cataract** is to
   a. Psychologist.
   b. Ophthalmologist.
   c. Otolaryngologist
   d. Otologist.
   e. Orthopedist.

6. **Jump** is to **surprise** as
   a. Run is to walk.
   b. Spook is to scare.
   c. Chuckle is to joke.
   d. Hop is to bunny.
   e. Sadness is to cry.

7. **Knife** is to **slice** as **fork** is to
   a. Cut.
   b. Spear
   c. Mouth.
   d. Spoon.
   e. Eat.

8. **Eat** is to **ate** as **spin** is to
   a. Thread.
   b. Spinned.
   c. Spinning.
   d. Spun.
   e. Spins.

9. **Obviate** is to **preclude** as
   a. Exclude is to include.
   b. Conceal is to avert.
   c. Pontificate is to ponder.
   d. Appease is to placate.
   e. Ostentatious is to poignant.

10. **Acre** is to **area** as **fathom** is to
    a. Depth.
    b. Angle degree.
    c. Wind speed.
    d. Ship.
    e. Width.

11. **Green** is to **blue** as **orange** is to
    a. Red.
    b. Brown.
    c. Purple.
    d. Green.
    e. Blue.

12. **Peel** is to **orange** as
    a. Fur is to bear.
    b. Shed is to snake.
    c. Peel is to sunburn.
    d. Shell is to coconut.
    e. Fuzz is to peach.

13. **Sycophant** is to **flattery** as **raconteur** is to
    a. Heritage.
    b. Philosophies.
    c. Idioms.
    d. Artwork.
    e. Anecdotes.

14. **Viscous** is to **runny** as
    a. Somber is to merciful.
    b. Vacuous is to nostalgic.
    c. Plunder is to laud.
    d. River is to stream.
    e. Obscure is to unequivocal.

15. **Nadir** is to **zenith** as **valley** is to
    a. Depression.
    b. Pinnacle.
    c. Climb.
    d. Rise
    e. Slope.

16. **Pundit** is to **expertise** as **scholar** is to
    a. Study.
    b. Learning.
    c. Novice.
    d. Teacher.
    e. Erudition.

17. **Blueprint** is to **architect** as
    a. Stethoscope is to doctor.
    b. Model is to train.
    c. Lathe is to craftsman.
    d. Outline is to drawing.
    e. Score is to composer.

18. **Evidence** is to **detective** as **gold** is to
    a. Jeweler.
    b. Pan.
    c. Prospector.
    d. Magnet.
    e. Archaeologist.

19. **Odometer** is to **distance** as **caliper** is to
    a. Pressure.
    b. Thickness.
    c. Wind.
    d. Brake.
    e. Body.

20. **Radish** is to **vegetable** as
    a. Garbanzo is to legume.
    b. Pineapple is to berry.
    c. Lettuce is to spinach.
    d. Cucumber is to salad.
    e. Citrus is to fruit.

21. **Adore** is to **appreciate** as **loathe** is to
    a. Detest.
    b. Hate.
    c. Appreciate.
    d. Dislike.
    e. Fear.

22. **Thanksgiving** is to **November** as
    a. Summer is to vacation.
    b. Easter is to spring.
    c. Labor Day is to January.
    d. Holiday is to celebration.
    e. Christmas is to December.

23. **Alligator** is to **reptile** as **elephant** is to
   a. Mammal.
   b. Animal.
   c. Australia.
   d. Marsupial.
   e. Bear.

24. **Nylon** is to **parachute** as
   a. Plexiglass is to glass.
   b. Neoprene is to wetsuit.
   c. Sweater is to wool.
   d. Wood is to bark.
   e. Ice is to cube.

25. **Painter** is to **easel** as **weaver** is to
   a. Pattern.
   b. Tapestry.
   c. Yarn.
   d. Needle.
   e. Loom.

26. **Coarse** is to **rough** as
   a. Smooth is to grainy.
   b. Butter is to sandpaper.
   c. Funny is to amusing.
   d. Hot is to ocean.
   e. Dinner is to breakfast.

27. **Fluctuate** is to **persist** as
   a. Apple is to orange.
   b. Guitar is to instrument.
   c. Tail is to cat.
   d. Tremble is to shake.
   e. Happy is to sad.

28. **Engine** is to **car** as
   a. Lemon is to fruit.
   b. Branch is to tree.
   c. Brave is to courageous.
   d. Rubber is to cement.
   e. Silverware is to cutlery.

29. **Lotion** is to **hydrate** as
   a. Plant is to flower.
   b. Angry is to irate.
   c. Ear is to head.
   d. Pot is to boil.
   e. Fumigate is to vaporize.

30. **Green** is to **go** as

a. Hurricane is to tornado.
b. Yellow is to color.
c. Crispy is to chewy.
d. Dove is to peace.
e. Salamander is to snake.

Reading Comprehension

Read the statement or passage and then choose the best answer to the question. Answer the question based on what is stated or implied in the passage.

1. Rehabilitation rather than punitive justice is becoming much more popular in prisons around the world. Prisons in America, especially, where the recidivism rate is 67 percent, would benefit from mimicking prison tactics in Norway, which has a recidivism rate of only 20 percent. In Norway, the idea is that a rehabilitated prisoner is much less likely to offend than one harshly punished. Rehabilitation includes proper treatment for substance abuse, psychotherapy, healthcare and dental care, and education programs.

Which of the following best captures the author's purpose?

a. To show the audience one of the effects of criminal rehabilitation by comparison
b. To persuade the audience to donate to American prisons for education programs
c. To convince the audience of the harsh conditions of American prisons
d. To inform the audience of the incredibly lax system of Norway prisons

2. What a lark! What a plunge! For so it had always seemed to her, when, with a little squeak of the hinges, which she could hear now, she had burst open the French windows and plunged at Bourton into the open air. How fresh, how calm, stiller than this of course, the air was in the early morning; like the flap of a wave; the kiss of a wave; chill and sharp and yet (for a girl of eighteen as she then was) solemn, feeling as she did, standing there at the open window, that something awful was about to happen; looking at the flowers, at the trees with the smoke winding off them and the rooks rising, falling; standing and looking until Peter Walsh said, "Musing among the vegetables?"— was that it?—"I prefer men to cauliflowers"— was that it? He must have said it at breakfast one morning when she had gone out on to the terrace— Peter Walsh. He would be back from India one of these days, June or July, she forgot which, for his letters were awfully dull; it was his sayings one remembered; his eyes, his pocket-knife, his smile, his grumpiness and, when millions of things had utterly vanished—how strange it was!—a few sayings like this about cabbages.

From *Mrs. Dalloway* by Virginia Woolf

What was the narrator feeling right before Peter Walsh's voice distracted her?

a. A spark of excitement for the morning.
b. Anger at the larks.
c. A sense of foreboding.
d. Confusion at the weather.

3. According to the plan of the convention, all judges who may be appointed by the United States are to hold their offices *during good behavior*, which is conformable to the most approved of the State constitutions and among the rest, to that of this State. Its propriety having been drawn into question by the adversaries of that plan, is no light symptom of the rage for objection, which disorders their imaginations and judgments. The standard of good behavior for the continuance in office of the judicial magistracy, is certainly one of the most valuable of the modern improvements in the practice of government. In a monarchy it is an excellent barrier to the despotism of the prince; in a republic it is a no less excellent barrier to the encroachments and oppressions of the representative body. And it is the best expedient which can be devised in any government, to secure a steady, upright, and impartial administration of the laws.

From *The Federalist No. 78* by Alexander Hamilton

What is Hamilton's point in this excerpt?

a. To show the audience that despotism within a monarchy is no longer the standard practice in the States.
b. To convince the audience that judges holding their positions based on good behavior is a practical way to avoid corruption.
c. To persuade the audience that having good behavior should be the primary characteristic of a person in a government body and their voting habits should reflect this.
d. To convey the position that judges who serve for a lifetime will not be perfect and therefore we must forgive them for their bad behavior when it arises.

4. There was a man named Webster lived in a town of twenty-five thousand people in the state of Wisconsin. He had a wife named Mary and a daughter named Jane and he was himself a fairly prosperous manufacturer of washing machines. When the thing happened of which I am about to write he was thirty-seven or thirty-eight years old and his one child, the daughter, was seventeen. Of the details of his life up to the time a certain revolution happened within him it will be unnecessary to speak. He was however a rather quiet man inclined to have dreams which he tried to crush out of himself in order that he function as a washing machine manufacturer; and no doubt, at odd moments, when he was on a train going some place or perhaps on Sunday afternoons in the summer when he went alone to the deserted office of the factory and sat several hours looking out at a window and along a railroad track, he gave way to dreams.

From *Many Marriages* by Sherwood Anderson

What does the author mean by the following sentence?

"Of the details of his life up to the time a certain revolution happened within him it will be unnecessary to speak."

a. The details of his external life don't matter; only the details of his internal life matter.
b. Whatever happened in his life before he had a certain internal change is irrelevant.
c. He had a traumatic experience earlier in his life which rendered it impossible for him to speak.
d. Before the revolution, he was a lighthearted man who always wished to speak to others no matter who they were.

5. The old castle soon proved to be too small for the family, and in September 1853 the foundation-stone of a new house was laid. After the ceremony the workmen were entertained at dinner, which was followed by Highland games and dancing in the ballroom.

Two years later they entered the new castle, which the Queen described as "charming; the rooms delightful; the furniture, papers, everything perfection."

The Prince was untiring in planning improvements, and in 1856 the Queen wrote: "Every year my heart becomes more fixed in this dear Paradise, and so much more so now, that *all* has become my dearest Albert's *own* creation, own work, own building, own laying out as at Osborne; and his great taste, and the impress of his dear hand, have been stamped everywhere. He was very busy today, settling and arranging many things for next year."

From the biography *Queen Victoria* by E. Gordon Browne, M.A.

What does the word *impress* mean in the third paragraph?

a. To affect strongly in feeling
b. To urge something to be done
c. A certain characteristic or quality imposed upon something else
d. To press a thing onto something else

6. Having completed these preparations, Mr. Booth entered the theater by the stage door; summoned one of the scene shifters, Mr. John Spangler, emerged through the same door with that individual, leaving the door open, and left the mare in his hands to be held until he (Booth) should return. Booth who was even more fashionably and richly dressed than usual, walked thence around to the front of the theater, and went in. Ascending to the dress circle, he stood for a little time gazing around upon the audience and occasionally upon the stage in his usual graceful manner. He was subsequently observed by Mr. Ford, the proprietor of the theater, to be slowly elbowing his way through the crowd that packed the rear of the dress circle toward the right side, at the extremity of which was the box where Mr. and Mrs. Lincoln and their companions were seated. Mr. Ford casually noticed this as a slightly extraordinary symptom of interest on the part of an actor so familiar with the routine of the theater and the play.

From *The Life, Crime, and Capture of John Wilkes Booth* by George Alfred Townsend

What does the author mean by the last two sentences?

a. Mr. Ford was suspicious of Booth and assumed he was making his way to Mr. Lincoln's box.
b. Mr. Ford assumed Booth's movement throughout the theater was due to being familiar with the theater.
c. Mr. Ford thought that Booth was making his way to the theater lounge to find his companions.
d. Mr. Ford thought that Booth was elbowing his way to the dressing room to get ready for the play.

7. When we study more carefully the effect upon the milk of the different species of bacteria found in the dairy, we find that there is a great variety of changes which they produce when they are allowed to grow in milk. The dairyman experiences many troubles with his milk. It sometimes curdles without becoming acid. Sometimes it becomes bitter, or acquires an unpleasant "tainted" taste, or, again, a "soapy" taste. Occasionally a dairyman finds his milk becoming slimy, instead of souring and curdling in the normal fashion. At such times, after a number of hours, the milk becomes so slimy that it can be drawn into long threads. Such an infection proves very troublesome, for many a time it persists in spite of all attempts made to remedy it. Again, in other cases the milk will turn blue, acquiring about the time it becomes sour a beautiful sky-blue colour. Or it may become red, or occasionally yellow. All of these troubles the dairyman owes to the presence in his milk of unusual species of bacteria which grow there abundantly.

From *The Story of Germ Life* by Herbert William Conn

What is the tone of this passage?

a. Excitement
b. Anger
c. Neutral
d. Sorrowful

8. Portland is a very beautiful city of 60,000 inhabitants, and situated on the Willamette river twelve miles from its junction with the Columbia. It is perhaps true of many of the growing cities of the West, that they do not offer the same social advantages as the older cities of the East. But this is principally the case as to what may be called boom cities, where the larger part of the population is of that floating class which follows in the line of temporary growth for the purposes of speculation, and in no sense applies to those centers of trade whose prosperity is based on the solid foundation of legitimate business. As the metropolis of a vast section of country, having broad agricultural valleys filled with improved farms, surrounded by mountains rich in mineral wealth, and boundless forests of as fine timber as the world produces, the cause of Portland's growth and prosperity is the trade which it has as the center of collection and distribution of this great wealth of natural resources, and it has attracted, not the boomer and speculator, who find their profits in the wild excitement of the boom, but the merchant, manufacturer, and investor, who seek the surer if slower channels of legitimate business and investment. These have come from the East, most of them within the last few years. They came as seeking a better and wider field to engage in the same occupations they had followed in their Eastern homes, and bringing with them all the love of polite life which they had acquired there, have established here a new society, equaling in all respects that which they left behind. Here are as fine churches, as complete a system of schools, as fine residences, as great a love of music and art, as can be found at any city of the East of equal size.

From *Oregon, Washington, and Alaska. Sights and Scenes for the Tourist*, written by E.L. Lomax in 1890

What is a characteristic of a "boom city," as indicated by the passage?

a. A city that is built on solid business foundation of mineral wealth and farming.
b. An area of land on the west coast that quickly becomes populated by residents from the east coast.
c. A city that, due to the hot weather and dry climate, catches fire frequently, resulting in a devastating population drop.
d. A city whose population is made up of people who seek quick fortunes rather than building a solid business foundation.

9. The other of the minor deities at Nemi was Virbius. Legend had it that Virbius was the young Greek hero Hippolytus, chaste and fair, who learned the art of venery from the centaur Chiron, and spent all his days in the greenwood chasing wild beasts with the virgin huntress Artemis (the Greek counterpart of Diana) for his only comrade.

From *The Golden Bough* by Sir James George Frazer

Based on a prior knowledge of literature, the reader can infer this passage is taken from which of the following?

a. A eulogy
b. A myth
c. A historical document
d. A technical document

10. When I wrote the following passages, or rather the bulk of them, I lived alone, in the woods, a mile from any neighbor, in a house which I had built myself on the shore of Walden Pond, in Concord, Massachusetts, and earned my living by the labor of my hands only. I lived there two years and two months. At present I am a sojourner in civilized life again.

From *Walden* by Henry David Thoreau

What does the word *sojourner* most likely mean at the end of the passage?

a. Illegal immigrant
b. Temporary resident
c. Lifetime partner
d. Farm crop

11. I do not believe there are as many as five examples of deviation from the literalness of the text. Once only, I believe, have I transposed two lines for convenience of translation; the other deviations are (*if* they are such) a substitution of an *and* for a comma in order to make now and then the reading of a line musical. With these exceptions, I have sacrificed *everything* to faithfulness of rendering. My object was to make Pushkin himself, without a prompter, speak to English readers. To make him thus speak in a foreign tongue was indeed to place him at a disadvantage; and music and rhythm and harmony are indeed fine things, but truth is finer still. I wished to present not what Pushkin would have said, or should have said, if he had written in English, but what he does say in Russian. That, stripped from all ornament of his wonderful melody and grace of form, as he is in a translation, he still, even in the hard English tongue, soothes and stirs, is in itself a sign that through the individual soul of Pushkin sings that universal soul whose strains appeal forever to man, in whatever clime, under whatever sky.

From preface for *Poems by Alexander Pushkin* by Ivan Panin

According to the author, what is the most important aim of translation work?

a. To retain the beauty of the work.
b. To retain the truth of the work.
c. To retain the melody of the work.
d. To retain the form of the work.

12. Peach (Amygdalus persica)—In the last chapter I gave two cases of a peach-almond and a double-flowered almond which suddenly produced fruit closely resembling true peaches. I have also given many cases of peach-trees producing buds, which, when developed into branches, have yielded nectarines. We have seen that no less than six named and several unnamed varieties of the peach have thus produced several varieties of nectarine. I have shown that it is highly improbable that all these peach-trees, some of which are old varieties, and have been propagated by the million, are hybrids from the peach and nectarine, and that it is opposed to all analogy to attribute the occasional production of nectarines on peach-trees to the direct action of pollen from some neighbouring nectarine-tree. Several of the cases are highly remarkable, because, firstly, the fruit thus produced has sometimes been in part a nectarine and in part a peach; secondly, because nectarines thus suddenly produced have reproduced themselves by seed; and thirdly, because nectarines are produced from peach-trees from seed as well as from buds. The seed of the nectarine, on the other hand, occasionally produces peaches; and we have seen in one instance that a nectarine-tree yielded peaches by bud-variation. As the peach is certainly the oldest or primary variety, the production of peaches from nectarines, either by seeds or buds, may perhaps be considered as a case of reversion. Certain trees have also been described as indifferently bearing peaches or nectarines, and this may be considered as bud-variation carried to an extreme degree.

From *Variation of Animals and Plants* by Charles Darwin

Which of the following statements is NOT a detail from the passage?

a. At least six named varieties of the peach have produced several varieties of nectarine.
b. It is not probable that all of the peach trees mentioned are hybrids from the peach and nectarine.
c. An unremarkable case is the fact that nectarines are produced from peach trees from seeds as well as from buds.
d. The production of peaches from nectarines might be considered a case of reversion.

13. Which of the following is an accurate paraphrasing of the following sentence?

Certain trees have also been described as indifferently bearing peaches or nectarines, and this may be considered as bud-variation carried to an extreme degree.

a. Some trees are described as bearing peaches, and some trees have been described as bearing nectarines, but individually, the buds are extreme examples of variation.
b. One way in which bud variation is said to be carried to an extreme degree is when specific trees have been shown to casually produce peaches or nectarines.
c. Certain trees are indifferent to bud variation, as recently shown in the trees that produce both peaches and nectarines in the same season.
d. Nectarines and peaches are known to have cross-variation in their buds, which indifferently bear other sorts of fruit to an extreme degree.

14. Meanwhile the fog and darkness thickened so, that people ran about with flaring links, proffering their services to go before horses in carriages, and conduct them on their way. The ancient tower of a church, whose gruff old bell was always peeping slyly down at Scrooge out of a Gothic window in the wall, became invisible, and struck the hours and quarters in the clouds, with tremulous vibrations afterwards as if its teeth were chattering in its frozen head up there. The cold became intense. In the main street, at the corner of the court, some labourers were repairing the gas-pipes, and had lighted a great fire in a brazier, round which a party of ragged men and boys were gathered: warming their hands and winking their eyes before the blaze in rapture. The water-plug being left in solitude, its overflowings sullenly congealed, and turned to misanthropic ice. The brightness of the shops where holly sprigs and berries crackled in the lamp heat of the windows, made pale faces ruddy as they passed. Poulterers' and grocers' trades became a splendid joke; a glorious pageant, with which it was next to impossible to believe that such dull principles as bargain and sale had anything to do. The Lord Mayor, in the stronghold of the mighty Mansion House, gave orders to his fifty cooks and butlers to keep Christmas as a Lord Mayor's household should; and even the little tailor, whom he had fined five shillings on the previous Monday for being drunk and bloodthirsty in the streets, stirred up to-morrow's pudding in his garret, while his lean wife and the baby sallied out to buy the beef.

from *A Christmas Carol* by Charles Dickens

Which of the following can NOT be inferred from the passage?

a. The season of this narrative is in the wintertime.
b. The majority of the narrative is located in a bustling city street.
c. This passage takes place during the nighttime.
d. The Lord Mayor is a wealthy person within the narrative.

15. What does the author mean by the following sentence?
The brightness of the shops where holly sprigs and berries crackled in the lamp heat of the windows, made pale faces ruddy as they passed.

a. When people walked past the shops, their faces turned red because of the lamps in the windows that were also lighting up holly sprigs and berries.
b. Compared with the holly sprigs and berries and their crackling lamplight, everyone's face looked old when they walked by the shops.
c. When people walked past the shops, their faces looked cold and blue compared with the warm light of the shops, which were making the holly sprigs and berries glow.
d. While shop owners were cooking their holly sprigs and berries in the warm glow of the fire, people's faces lit up with excitement as they passed.

Questions 16–18 are based on the following passage from the book On the Trail *by Lina Beard and Adelia Belle Beard:*

For any journey, by rail or by boat, one has a general idea of the direction to be taken, the character of the land or water to be crossed, and of what one will find at the end. So it should be in striking the trail. Learn all you can about the path you are to follow. Whether it is plain or obscure, wet or dry; where it leads; and its length, measured more by time than by actual miles. A smooth, even trail of five miles will not consume the time and strength that must be expended upon a trail of half that length which leads over uneven ground, varied by bogs and obstructed by rocks and fallen trees, or a trail that is all up-hill climbing. If you are a novice and accustomed to walking only over smooth and level ground, you must allow more time for covering the distance than an experienced person would require and must count upon the expenditure of more strength, because your feet are not trained to the wilderness paths with their pitfalls and traps for the unwary, and every nerve and muscle will be strained to secure a safe foothold amid the tangled roots, on the slippery, moss-covered logs, over precipitous rocks that lie in your path. It will take time to pick your way over boggy places where the water oozes up through the thin, loamy soil as through a sponge; and experience alone will teach you which hummock of grass or moss will make a safe stepping-place and will not sink beneath your weight and soak your feet with hidden water. Do not scorn to learn all you can about the trail you are to take . . . It is not that you hesitate to encounter difficulties, but that you may prepare for them. In unknown regions take a responsible guide with you, unless the trail is short, easily followed, and a frequented one. Do not go alone through lonely places; and, being on the trail, keep it and try no explorations of your own, at least not until you are quite familiar with the country and the ways of the wild.

16. What is the meaning of the word *novice* in this passage?
    a. Expert
    b. Beginner
    c. Child
    d. Adult

17. What does the author say about unknown regions?
    a. You should try and explore unknown regions to learn the land better.
    b. Unless the trail is short or frequented, you should take a responsible guide with you.
    c. All unknown regions will contain pitfalls, traps, and boggy places.
    d. It's better to travel unknown regions by rail than by foot.

18. Which statement is NOT a detail from the passage?
    a. Learning about the trail beforehand is imperative.
    b. Time will differ depending on the land.
    c. Once you are familiar with the outdoors, you can go places on your own.
    d. Be careful of wild animals on the trail you are on.

Questions 19 and 20 are based on the following passage from the novel Frankenstein *by Mary Wollstonecraft Shelley:*

I trembled excessively; I could not endure to think of, and far less to allude to, the occurrences of the preceding night. I walked with a quick pace, and we soon arrived at my college. I then reflected, and the thought made me shiver, that the creature whom I had left in my apartment might still be there, alive, and walking about. I dreaded to behold this monster; but I feared still more that Henry should see him. Entreating him, therefore, to remain a few minutes at the bottom of the stairs, I darted up towards my own room. My hand was already on the lock of the door before I recollected myself. I then paused; and a cold shivering came over me. I threw the door forcibly open, as children are accustomed to do when they expect a spectre to stand in waiting for them on the other side; but nothing appeared. I stepped fearfully in: the apartment was empty; and my bed-room was also freed from its hideous guest. I could hardly believe that so great a good fortune could have befallen me; but when I became assured that my enemy had indeed fled, I clapped my hands for joy, and ran down to Clerval.

19. Which statement is NOT a detail from the passage?
    a. The speaker trembles when thinking about the occurrences the night before.
    b. The speaker throws his own door open in a flurry.
    c. The speaker claps his hands for joy and runs down to his friend.
    d. The speaker sees a ghost in his home when he opens the door.

20. Which of the following sentences best summarizes the passage?
    a. The speaker meets his friend at an abandoned building where they attempt to hunt a creature that terrorized them the night before.
    b. The speaker and a friend arrive at the apartment where, to the speaker's relief, a terrifying creature from the night before appears to have left the space.
    c. The speaker is angry at his friend for letting loose a terrifying creature in his house, only to find the creature gone.
    d. The speaker is eager to show his friend a terrifying creature locked inside his apartment, only to find that the creature has disappeared from the apartment.

Clerical

1. Which word is a synonym for the word *facetious*?
    a. Important
    b. Rebellious
    c. Benign
    d. Flippant

2. Which word is misspelled?
    a. Poignent
    b. Reciprocity
    c. Bourgeois
    d. Ramification

3. Which number sequence is correct?
   a. 21 43 39 17
   b. 21 18 15 12
   c. 31 34 52 47
   d. 16 21 19 33

4. Solve the following problem: $(6 - 2)(9 \div 3)$.
   a. 13
   b. 6
   c. 12
   d. 7

5. Find the exact copy of the following code: XRD7903518&
   a. XR7D035981&
   b. XRD7903518&
   c. XRD&7803519
   d. XDR7903158&

6. Which word means the same as *regressive*?
   a. Deteriorating
   b. Relapse
   c. Digress
   d. Imminent

7. Which word is an antonym for *discord*?
   a. Agreement
   b. Clamor
   c. Tumult
   d. Dissonance

8. Solve the following problem: $770\ x\ 139\ =$
   a. 11,793
   b. 9,053
   c. 107,136
   d. 107,030

9. Solve the following problem: $7\ x\ 7\ x\ 7\ x\ 7\ x\ 7\ =$
   a. 16,807
   b. 16,881
   c. 23,435
   d. 11,049

10. Steven has eight apples. He eats two of them. Helen has nine apples. She gives half of them to Steven. How many apples does Steven have?
   a. 4.5
   b. 10.5
   c. 7
   d. 6.75

11. Which word is a synonym for the word *meticulous*?
   a. Trite
   b. Sloppy
   c. Stunted
   d. Careful

12. Which word is an antonym for the word *hostile*?
   a. Receptive
   b. Aggressive
   c. Inimical
   d. Intimidating

13. April works at both the mini mart and the drug store part time. How much total weekly income does she earn?

Stores	Hourly Wages	Weekly Hours Worked
Mini Mart	$11	Part time: 25 Full time: 40
Drug Store	$15	Part time: 16 Full time: 32
Supercenter	$22	Part time: 30 Full time: 40

   a. $525
   b. $475
   c. $515
   d. $680

14. Please select the exact copy of the following address from the list below: 23791 Maple Court Dr.
   a. 27391 Maple Court Dr.
   b. 23791 Mople Court Dr.
   c. 23791 Maple Court Dr.
   d. 23917 Maple Court Dr.

15. Which word means the opposite of *succinct*?
   a. Concise
   b. Verbose
   c. Brief
   d. Pithy

16. The names Amber Greene, Ashley Gregson, **Avery Greenly**, and Aaron Greily need to be arranged in alphabetical order by last name. In which position would the name in bold, Avery Greenly, be placed?
   a. First
   b. Second
   c. Third
   d. Fourth

17. According to the following table, which person would be coded 3-X-F?

Race	
African American	1
White	2
Asian American	3
Hispanic	4
Sex	
Male	X
Female	Y
Nonbinary	Z
Height	
4′9″-5′4″	D
5′5″-5′9″	E
5′10″-6′3″	F

a. An African American woman who is 5′7″
b. A nonbinary Asian American person who is 6′2″
c. An Asian American man who is 5′11″
d. A Caucasian woman who is 5′1″

18. What would be the code for a police officer with an associate's degree and 13 years of experience?

Occupation	
Teacher	A
Plumber	B
Police Officer	C
Flight Attendant	D
Years of Experience	
0-5	4
5-11	5
12-17	6
Education	
High School Diploma	L
Associates Degree	M
4-year College Degree	O

a. C-5-M
b. L-6-A
c. O-4-C
d. C-M-6

19. Place the following numbers in descending order.
1) 9 ¾
2) 9 ½
3) 10.5
4) 11.1
5) 8 $^{5}/_{8}$

a. 2, 1, 5, 3, 4
b. 5, 2, 1, 3, 4
c. 5, 1, 3, 2, 4
d. 4, 3, 1, 2, 5

20. What is another word for *prescient*?
a. Perceptive
b. Ignorant
c. Unknowing
d. Blatant

21. Solve the following problem: $397\ x\ 863\ =$
   a. 342,671
   b. 342,611
   c. 432,609
   d. 249,611

22. Find the exact copy of the following code: HRC7816$3L0
   a. HCR7816$31O
   b. HIC7816$3L0
   c. HRC7816$3L0
   d. HRC6817$3L0

23. Which word is misspelled?
   a. Prurient
   b. Cartharsys
   c. Visionary
   d. Treble

24. Which number should replace the pound sign?

14	#	686
4,802	33,614	235,298

   a. 98
   b. 42
   c. 70
   d. 84

25. What is the next number in the following sequence? 9, 18, 36, 72, 144...
   a. 216
   b. 312
   c. 576
   d. 288

26. There are 2,000 students at the University of Lynchburg. Of these students, 500 are studying nursing, 35 are studying film, 1,050 are studying the communication studies, and 415 are studying business. What percentage of students are studying film?
   a.15.75%
   b. 17.5%
   c. 1.75%
   d. 21%

27. Which word is an antonym for *malignant*?
   a. Wicked
   b. Malicious
   c. Malevolent
   d. Benevolent

28. Which word is a synonym for *respectively*?
    a. Arbitrarily
    b. Inordinately
    c. Separately
    d. Incommensurately

29. Find the exact copy of the following code: ^**#DL7634IIX
    a. ^**#DL7634IIX
    b. ^^#*DL7634IIX
    c. ^**#DL7634IX
    d. ^**#DL7643IIX

30. Solve the following problem: $1{,}444 \div 90 =$
    a. 12.5
    b. 16.04
    c. 17
    d. 21.7

31. Which word is spelled correctly?
    a. Bovine
    b. Definat
    c. Superbe
    d. Thorought

32. Which word is a synonym for *retired*?
    a. Present
    b. Incorporated
    c. Emeritus
    d. Unreserved

33. Solve the following problem: $350\ x\ 79 =$
    a. 1,479
    b. 27,650
    c. 21,650
    d. 7,961

34. Arrange the following names in *descending* alphabetical order: Leonard, Linwood, Lewis, Louis, Lucas.
    a. Leonard, Linwood, Lewis, Louis, and Lucas
    b. Lucas, Louis, Lewis, Linwood, Leonard
    c. Lucas, Louis, Linwood, Lewis, Leonard
    d. Leonard, Lewis, Linwood, Louis, Lucas

35. Solve the following problem: 815 + 369 =
    a. 1,104
    b. 1,284
    c. 1,286
    d. 1,184

36. Rick plans to study for his exam for 21 hours each week. He plans to spend 3 hours per night studying. So far, Rick has studied for 4 nights this week. How many more hours of study time are left this week?

a. 9
b. 6
c. 7
d. 8

Use the following chart for the next three questions.

State	Most Populous City	Population
Virginia	Virginia Beach	449,974
Rhode Island	Providence	179,883
Kansas	Wichita	389,938
California	Los Angeles	3,979,576
Missouri	Kansas City	495,327

37. How many cities have a population higher than 450, 000?
a. Three
b. Two
c. One
d. None

38. What is the average population of Providence and Wichita?
a. 284,910
b. 359,284
c. 256,632
d.135,789

39. Which state has the second most populous city on the chart?
a. Missouri
b. California
c. Virginia
d. Kansas

40. What is $13^{2?}$
a. 26
b. 39
c.169
d.196

41. Which number should replace the pound sign?

128	#	32
16	8	4

a. 12
b. 64
c. 24
d. 48

42. Which word is a synonym for the word *evade*?

a. Confront
b. Consume
c. Validate
d. Skirt

43. Which word is an antonym for the word *indictment*?

a. Complaint
b. Charge
c. Pardon
d. Warrant

44. How should the following sentence be written?
While your out, pick up you're clothes from the dry cleaner.

a. While your out, pick up your clothes from the dry cleaner.
b. While you're out, pick up your clothes from the dry cleaner.
c. The sentence is correct as written.
d. While you're out, pick up you're clothes from the dry cleaner.

45. How should the following sentence be written? More than one choice may be selected.
Walter is at the top of his class, he will be valedictorian.

a. The sentence is correct as written.
b. Walter is at the top of his class... he will be valedictorian.
c. Walter is at the top of his class, and he will be valedictorian.
d. Walter is at the top of his class. He will be valedictorian.

46. What is 3^9?

a. 19,683
b. 333
c. 3,639
d. 27

47. What is a synonym for the word *adverse*?

a. Suitable
b. Unfavorable
c. Minute
d. Cowardly

48. Arrange the following names in alphabetical order: Richard, Rick, Ryan, Riley, Regan.
  a. Regan, Rick, Richard, Riley, Ryan
  b. Ryan, Regan, Richard, Rick, Riley,
  c. Regan, Richard, Rick, Riley, Ryan
  d. Riley, Regan, Rick, Richard, Ryan

49. Solve the following problem: $1683 - 765 =$
  a. 918
  b. 1023
  c. 809
  d. 716

50. Find the exact copy of the following code: ALSX13390V
  a. ALSX13330V
  b. ASLX3390V
  c. ALXS19330V
  d. ALSX13390V

Mathematics

1. What is $\frac{12}{60}$ converted to a percentage?
  a. 0.20
  b. 20%
  c. 25%
  d. 12%

2. Which of the following represents the correct sum of $\frac{14}{15}$ and $\frac{2}{5}$, in lowest possible terms?
  a. $\frac{20}{15}$
  b. $\frac{4}{3}$
  c. $\frac{16}{20}$
  d. $\frac{4}{5}$

3. What is the product of $\frac{5}{14}$ and $\frac{7}{20}$, in lowest possible terms?
  a. $\frac{1}{8}$
  b. $\frac{35}{280}$
  c. $\frac{12}{34}$
  d. $\frac{1}{2}$

4. What is the result of dividing 24 by $\frac{8}{5}$, in lowest possible terms?

a. $\frac{5}{3}$

b. $\frac{3}{5}$

c. $\frac{120}{8}$

d. 15

5. Subtract $\frac{5}{14}$ from $\frac{5}{24}$. Which of the following is the correct result?

a. $\frac{25}{168}$

b. 0

c. $-\frac{25}{168}$

d. $\frac{1}{10}$

6. Which of the following is a correct mathematical statement?

a. $\frac{1}{3} < -\frac{4}{3}$

b. $-\frac{1}{3} > \frac{4}{3}$

c. $\frac{1}{3} > -\frac{4}{3}$

d. $-\frac{1}{3} \geq \frac{4}{3}$

7. Which of the following is INCORRECT?

a. $-\frac{1}{5} < \frac{4}{5}$

b. $\frac{4}{5} > -\frac{1}{5}$

c. $-\frac{1}{5} > \frac{4}{5}$

d. $\frac{1}{5} > -\frac{4}{5}$

8. How many cases of cola can Lexi purchase if each case is $3.50 and she has $40?

a. 10
b. 12
c. 11.4
d. 11

9. A car manufacturer usually makes 15,412 SUVs, 25,815 station wagons, 50,412 sedans, 8,123 trucks, and 18,312 hybrids a month. About how many cars are manufactured each month?

a. 120,000
b. 200,000
c. 300,000
d. 12,000

10. Each year, a family goes to the grocery store every week and spends $105. About how much does the family spend annually on groceries?

a. $10,000
b. $50,000
c. $500
d. $5,000

11. A grocery store sold 48 bags of apples in one day, and 9 of the bags contained Granny Smith apples. The rest contained Red Delicious apples. What is the ratio of bags of Granny Smith to bags of Red Delicious apples that were sold?

a. 48:9
b. 39:9
c. 9:48
d. 9:39

12. If Oscar's bank account totaled $4,000 in March and $4,900 in June, what was the rate of change in his bank account over those three months?

a. $900 a month
b. $300 a month
c. $4,900 a month
d. $100 a month

13. Erin and Katie work at the same ice cream shop. Together, they always work less than 21 hours a week. In a week, if Katie worked two times as many hours as Erin, how many hours did Erin work?

a. Less than 7 hours
b. Less than or equal to 7 hours
c. More than 7 hours
d. Less than 8 hours

14. Which of the following is the correct decimal form of the fraction $\frac{14}{33}$ rounded to the nearest hundredth place?

a. 0.420
b. 0.42
c. 0.424
d. 0.140

15. Gina took an algebra test last Friday. There were 35 questions, and she answered 60% of them correctly. How many correct answers did she have?

a. 35
b. 20
c. 21
d. 25

16. Paul took a written driving test, and he got 12 of the questions correct. If he answered 75% of the questions correctly, how many problems were there in the test?

a. 25
b. 16
c. 20
d. 18

17. What is the solution to the equation $3(x + 2) = 14x - 5$?

a. $x = 1$

b. $x = 0$

c. All real numbers

d. There is no solution

18. What is the solution to the equation $10 - 5x + 2 = 7x + 12 - 12x$?

a. $x = 1$

b. $x = 0$

c. All real numbers

d. There is no solution

19. Which of the following is the result when solving the equation $4(x + 5) + 6 = 2(2x + 3)$?

a. $x = 26$

b. $x = 6$

c. All real numbers

d. There is no solution

20. Two consecutive integers exist such that the sum of three times the first and two less than the second is equal to 411. What are those integers?

a. 103 and 104

b. 104 and 105

c. 102 and 103

d. 100 and 101

21. In a neighborhood, 15 out of 80 of the households have children under the age of 18. What percentage of the households have children?

a. 0.1875%

b. 18.75%

c. 1.875%

d. 15%

22. If a car is purchased for $15,395 with a 7.25% sales tax, what is the total price?

a. $15,395.07

b. $16,511.14

c. $16,411.13

d. $15,402

23. From the chart below, which two are preferred by more men than women?

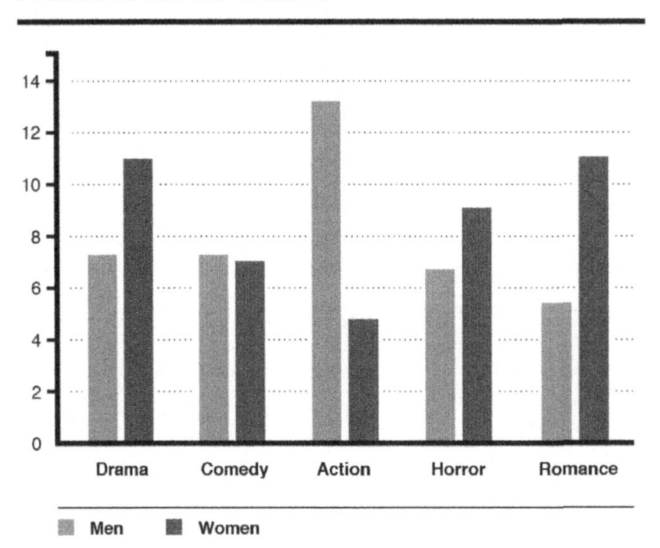

a. Comedy and Action
b. Drama and Comedy
c. Action and Horror
d. Action and Romance

24. Which type of graph best represents a continuous change over a period of time?
a. Bar graph
b. Line graph
c. Pie graph
d. Histogram

25. Using the graph below, what is the mean number of visitors for the first 4 hours?

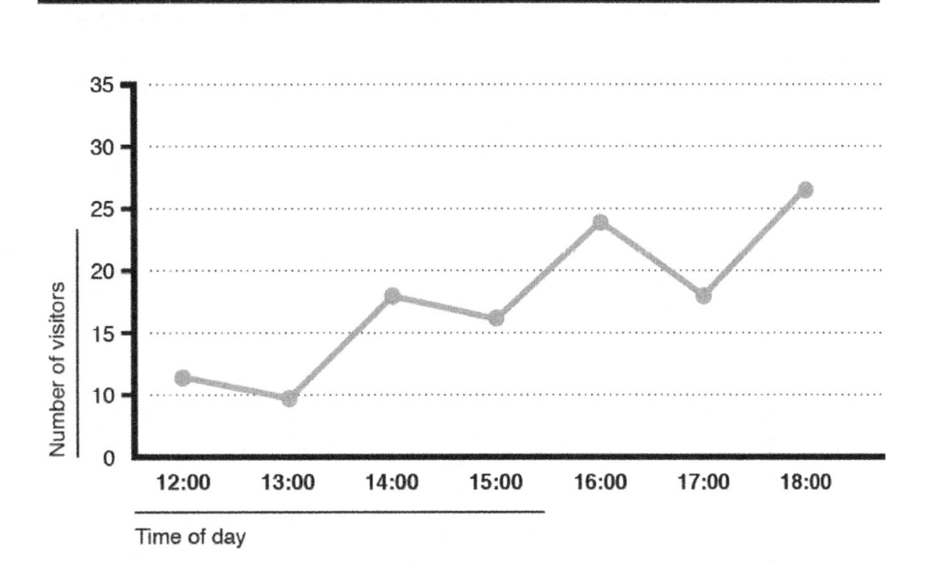

a. 12
b. 13
c. 14
d. 15

26. What is the mode for the grades shown in the chart below?

Science Grades	
Jerry	65
Bill	95
Anna	80
Beth	95
Sara	85
Ben	72
Jordan	98

a. 65
b. 33
c. 95
d. 90

27. What type of relationship is there between age and attention span as represented in the graph below?

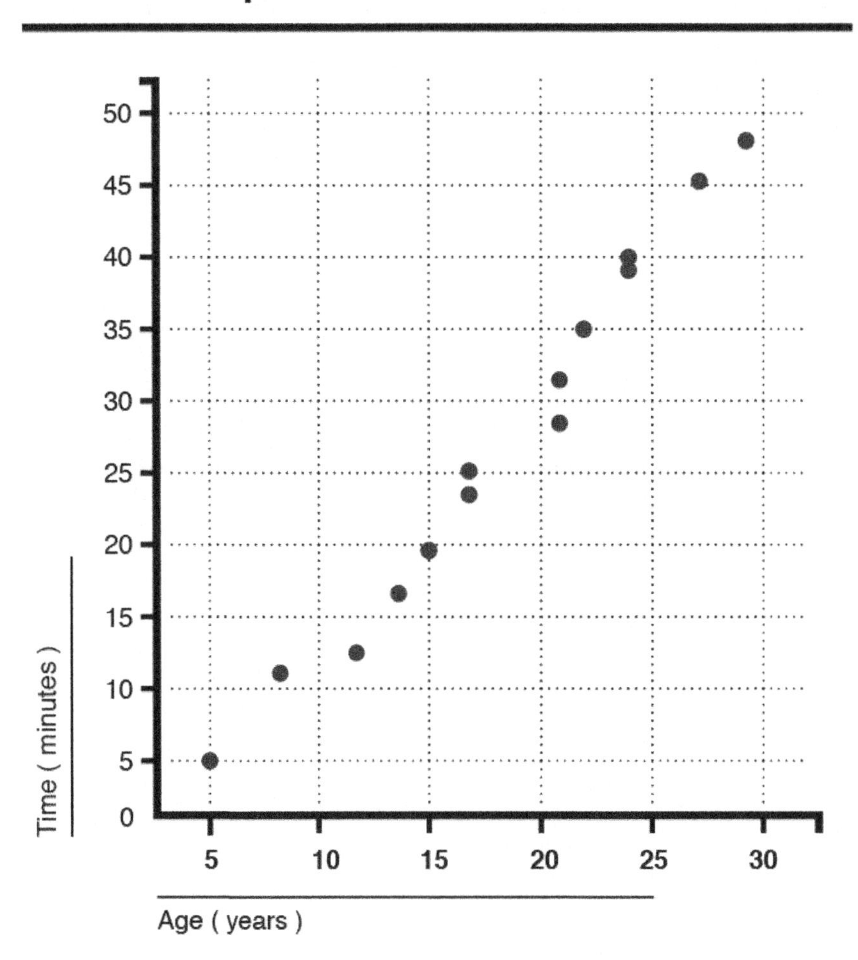

a. No correlation
b. Positive correlation
c. Negative correlation
d. Weak correlation

28. How many kiloliters are in 6 liters?
a. 6,000
b. 600
c. 0.006
d. 0.0006

29. Which of the following relations is a function?
a. {(1, 4), (1, 3), (2, 4), (5, 6)}
b. {(-1, -1), (-2, -2), (-3, -3), (-4, -4)}
c. {(0, 0), (1, 0), (2, 0), (1, 1)}
d. {(1, 0), (1, 2), (1, 3), (1, 4)}

30. Find the indicated function value: $f(5)$ for $f(x) = x^2 - 2x + 1$.

a. 16
b. 1
c. 5
d. Does not exist

31. What is the domain of $f(x) = 4x^2 + 2x - 1$?

a. $(0, \infty)$
b. $(-\infty, 0)$
c. $(-\infty, \infty)$
d. $(-1, 4)$

32. What is the range of the polynomial function $f(x) = 2x^2 + 5$?

a. $(-\infty, \infty)$
b. $(2, \infty)$
c. $(0, \infty)$
d. $[5, \infty)$

33. For which two values of x is $g(x) = 4x + 4$ equal to $g(x) = x^2 + 3x + 2$?

a. 1, 0
b. 1, -2
c. -1, 2
d. 1, 2

34. The population of coyotes in the local national forest has been declining since 2000. The population can be modeled by the function $y = -(x - 2)^2 + 1600$, where y represents number of coyotes and x represents the number of years past 2000. When will there be no more coyotes?

a. 2020
b. 2040
c. 2012
d. 2042

35. A ball is thrown up from a building that is 800 feet high. Its position s in feet above the ground is given by the function $s = -32t^2 + 90t + 800$, where t is the number of seconds since the ball was thrown. How long will it take for the ball to come back to its starting point? Round your answer to the nearest tenth of a second.

a. 0 seconds
b. 2.8 seconds
c. 3 seconds
d. 8 seconds

36. What is the domain of the following rational function?

$$f(x) = \frac{x^3 + 2x + 1}{2 - x}$$

a. $(-\infty, -2) \cup (-2, \infty)$
b. $(-\infty, 2) \cup (2, \infty)$
c. $(2, \infty)$
d. $(-2, \infty)$

37. What is the missing length *x*?

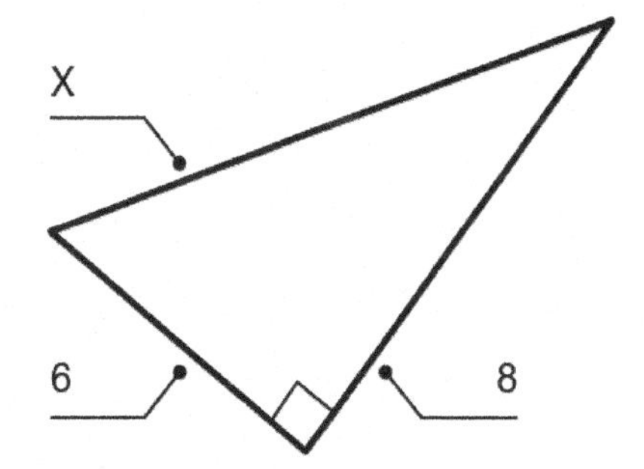

a. 6
b. 14
c. 10
d. 100

38. A study of adult drivers finds that it is likely that an adult driver wears his seatbelt. Which of the following could be the probability that an adult driver wears his seat belt?
a. 0.90
b. 0.05
c. 0.25
d. 0

39. What is the solution to the following linear inequality?

$$7 - \frac{4}{5}x < \frac{3}{5}$$

a. $(-\infty, 8)$
b. $(8, \infty)$
c. $[8, \infty)$
d. $(-\infty, 8]$

40. What is the solution to the following system of linear equations?

$$2x + y = 14$$

$$4x + 2y = -28$$

a. (0, 0)
b. (14, -28)
c. All real numbers
d. There is no solution

41. Which of the following is perpendicular to the line $4x + 7y = 23$?
a. $y = -\frac{4}{7}x + 23$
b. $y = \frac{7}{4}x - 12$
c. $4x + 7y = 14$
d. $y = -\frac{7}{4}x + 11$

42. What is the solution to the following system of equations?

$$2x - y = 6$$

$$y = 8x$$

a. (1, 8)
b. (-1, 8)
c. (-1, -8)
d. There is no solution.

43. The mass of the moon is about 7.348×10^{22} kilograms and the mass of Earth is 5.972×10^{24} kilograms. How many times *greater* is Earth's mass than the moon's mass?

a. 8.127×10^{1}
b. 8.127
c. 812.7
d. 8.127×10^{-1}

44. The percentage of smokers above the age of 18 in 2000 was 23.2 percent. The percentage of smokers over the age of 18 in 2015 was 15.1 percent. Find the average rate of change in the percentage of smokers over the age of 18 from 2000 to 2015.

a. -.54 percent
b. -54 percent
c. -5.4 percent
d. -15 percent

45. Triple the difference of five and a number is equal to the sum of that number and 5. What is the number?

a. 5
b. 2
c. 5.5
d. 2.5

46. In order to estimate deer population in a forest, biologists obtained a sample of deer in that forest and tagged each one of them. The sample had 300 deer in total. They returned a week later and harmlessly captured 400 deer, and 5 were tagged. Using this information, which of the following is the best estimate of the total number of deer in the forest?

a. 24,000 deer
b. 30,000 deer
c. 40,000 deer
d. 100,000 deer

47. What is the correct factorization of the following binomial?

$$2y^3 - 128$$

a. $2(y + 8)(y - 8)$
b. $2(y - 4)(y^2 + 4y + 16)$
c. $2(y - 4)(y + 4)^2$
d. $2(y - 4)^3$

48. What is the simplified form of $(4y^3)^4(3y^7)^2$?

a. $12y^{26}$

b. $2304y^{16}$

c. $12y^{14}$

d. $2304y^{26}$

49. The number of members of the House of Representatives varies directly with the total population in a state. If the state of New York has 19,800,000 residents and has 27 total representatives, how many should Ohio have with a population of 11,800,000?

a. 10

b. 16

c. 11

d. 5

50. The following set represents the test scores from a university class: {35, 79, 80, 87, 87, 90, 92, 95, 95, 98, 99}. If the outlier is removed from this set, which of the following is TRUE?

a. The mean and the median will decrease.

b. The mean and the median will increase.

c. The mean and the mode will increase.

d. The mean and the mode will decrease.

Answer Explanations

Spelling

1. A: The word *cacofony* should be spelled *cacophony. Beige, apathy,* and *courage* are the correct spellings.

2. B: The word *antham* should be spelled *anthem. Temporary, statute,* and *reasonable* are spelled correctly.

3. B: The word *jaring* should be *jarring. Unpleasant, corporate,* and *roulette* are spelled correctly.

4. D: The word *sinphony* should be *symphony. Marred, cajole,* and *blanketed* are the correct spellings.

5. D: The word *transfered* should be *transferred. Statement, balance,* and *minimum* are spelled correctly.

6. A: *Blasphamy* is misspelled; the correct spelling is *blasphemy. Heretic, religious,* and *canon* are all correct.

7. C: The word *freightened* is incorrect; it should be spelled *frightened.* Courage, hopeful, and mistaken are all correct.

8. D: The word *managable* should be spelled *manageable. Difficult, intense,* and *ascetic* are spelled correctly.

9. B: *Combersome* is misspelled; it should be spelled *cumbersome. Urgent, deficit,* and *language* are all correct.

10. B: *Minute, stringent,* and *sedition* are spelled correctly. The word *caripace* should be spelled *carapace.*

11. A: *Hipocracy* should be spelled *hypocrisy.* The words *legend, dominion,* and *history* are correct.

12. A: The words *riptide, sacrament,* and *blatant* are spelled correctly. *Mileiu* should be spelled *milieu.*

13. C: The word *ambiant* should be spelled *ambient. Carriage, decorative,* and *misshapen* are spelled correctly.

14. D: *Despondant* should be spelled *despondent. Dangerous, decedent,* and *plagiarism* are correct.

15. B: The word *rambuncteous* should be spelled *rambunctious. Plaintive, banal,* and *finale* are spelled correctly.

16. A: *Longitude, climate,* and *category* are spelled correctly. The word *despotte* should be spelled *despot.*

17. B: *Laborious, tradesman,* and *transition* are correct. *Cateclysmic* should be spelled *cataclysmic.*

18. C: *Abhorent* should be spelled *abhorrent. Dissolution, contractual,* and *seismic* are spelled correctly.

19. A: The word *rudamentary* should be spelled *rudimentary. Advantageous, torpor,* and *tumultuous* are spelled correctly.

20. B: The word *empediment* should be spelled *impediment. Succession, seizure,* and *electricity* are correct.

21. D: *Sacrosanct, judicious,* and *topple* are correct. *Anthisis* should be spelled *antithesis.*

22. A: The word *judiscious* should be spelled *judicious. Eliminate, bourgeoisie,* and *exclamatory* are spelled correctly.

23. B: *Annoint* should be spelled *anoint. Incredulous, tableau,* and *missile* are correct.

24. D: The word *misscommunication* should be spelled *miscommunication. Ablelist, dexterous,* and *incognito* are correct.

25. B: *Archery, incongruous,* and *congressional* are spelled correctly. *Mathmetician* should be spelled *mathematician.*

26. C: The word *delicatesin* should be spelled *delicatessen. Magnanimous, turbulent,* and *ardor* are spelled correctly.

27. A: The word *detritous* should be spelled *detritus. Siphon, category,* and *relegate* are spelled correctly.

28. D: Riboflavin, mangy, and superlative are spelled correctly. The word *delerium* should be spelled *delirium.*

29. C: *Spontanaity* should be spelled *spontaneity. Podcast, dapper,* and *arduous* are spelled correctly.

30. A: The word *malecotent* should be spelled *malcontent. Execute, southerly,* and *catholic* are correct.

31. D: *Ambient, localized,* and *ammunition* are spelled correctly. *Vaxinated* should be spelled *vaccinated.*

32. A: The word *purplexed* should be spelled *perplexed.* The words *superfluous, candidate,* and *incandescent* are correct.

33. B: The word *eccological* should be spelled *ecological. Collapse, realization,* and *mundane* are spelled correctly.

34. C: *Economist, publication,* and *cartridge* are spelled correctly. The word *negocioble* should be spelled *negotiable.*

35. A: *Piutocrat* should be spelled *plutocrat. Cathartic, greenery,* and *hibiscus* are spelled correctly.

36. A: *Demonstretive* should be spelled *demonstrative. Artisanal, umbilical,* and *collaborative* are correct.

37. B: The word *tertiery* should be spelled *tertiary. Biological, robust,* and *endemic* are spelled correctly.

38. D: *Systemic, irrigation,* and *furtive* are spelled correctly. The word *etamology* should be spelled *etymology.*

39. C: The word *idealogical* should be spelled *ideological. Comprehensive, optimism,* and *iterative* are spelled correctly.

40. A: *Emmersion* should be spelled *immersion. Punitive, explanation,* and *transcription* are spelled correctly.

41. B: *Succumb*, *renewable*, and *credible* are correct. *Nefareous* should be spelled *nefarious*.

42. A: The word *incontravertible* should be spelled *incontrovertible*. Discussion, impurity, and disengagement are spelled correctly.

43. C: The word *approppo* should be spelled *apropos*. *Academia*, *workshop*, and *opportunities* are correct.

44. C: *Machination*, *revival*, and *acquisition* are spelled correctly. *Adversariel* should be spelled *adversarial*.

45. D: The word *lagastician* should be spelled *logistician*. *Development*, *acceptable*, and *training* are spelled correctly.

46. B: *Telocommute* should be spelled *telecommute*. *Connectivity*, *visualization*, and *consecrate* are spelled correctly.

47. A: The word *clairivoyant* should be spelled *clairvoyant*. *Clearance*, *xylophone*, and *zoonotic* are spelled correctly.

48. B: *Technocratic*, *corporate*, and *hexagon* are correct. *Retroffit* should be spelled *retrofit*.

49. D: The word *hydroplain* should be spelled *hydroplane*. *Leaflet*, *evident*, and *mercurial* are spelled correctly.

50. A: *Councelor* should be spelled *counselor*. *Reservation*, *buoy*, and *gargantuan* are correct.

Vocabulary

1. E: Someone who is *wary* is overly cautious or apprehensive. This word is often used in the context of being watchful or on guard about a potential danger.

2. E: *Proximity* is defined as closeness, or the state or quality of being near in place, time, or relation.

3. C: As an adjective, *parsimonious* means frugal to the point of being stingy, or very unwilling to spend money, which is similar to being miserly.

4. B: The noun *propriety* means suitability or appropriateness to the given circumstances or purpose. It can mean conformity to accepted standards, particularly as they relate to good behavior or manners. In this way, *propriety* can be considered to mean the state or quality of being proper.

5. C: A *boon* is a benefit, blessing, or something to be thankful for. In an alternative usage, it can be a favor or a benefit given upon request.

6. E: *Strife* is a noun that is defined as bitter or vigorous discord, conflict, or dissension. It can mean a fight or struggle, or other act of contention.

7. B: *Qualm* is a noun that means a feeling of apprehension or uneasiness. A girl who is just learning to ride a bike may have qualms about getting back on the saddle after taking a bad fall. It may also refer to an uneasy feeling related to one's conscience as it pertains to his or her actions. For example, a man with poor morals may have no qualms about lying on his tax return.

8. D: *Valor* is bravery or courage when facing a formidable danger. It often relates to strength of mind or spirit during battle or acting heroically in such situations.

9. C: *Zeal* is eagerness, fervor, or ardent desire in the pursuit of something. For example, a competitive collegiate baseball player's zeal to succeed in his sport may compromise his academic performance.

10: A: To *extol* is to highly praise, laud, or glorify. People often extol the achievements of their heroes or mentors.

11. B: To *reproach* means to blame, find fault, or severely criticize. It can also be defined as expressing significant disapproval. It is often used in the phrase "beyond reproach" as in, "her violin performance was beyond reproach." In this context, it means her playing was so good that it evaded any possibility of criticism.

12. D: *Milieu* refers to someone's cultural surroundings or environment.

13. E: *Guile* can be defined as the quality of being cunning or crafty and skilled in deception. Someone may use guile to trick or deceive someone.

14. A: As a verb, *assent* means to express agreement, give consent, or acquiesce. A job candidate might assent to an interviewer's request to perform a background check. As a noun, it means an agreement, acceptance, or acquiescence.

15. B: *Dearth* means a lack of something or a scarcity or shortage. For example, a local library might have a dearth of information pertaining to an esoteric topic.

16. B: *Conspicuous* means to be visually or mentally obvious. Something conspicuous stands out, is clearly visible, or may attract attention.

17. D: *Onerous* most closely means burdensome or troublesome. It usually is used to describe a task or obligation that may impose a hardship or burden, often which may be perceived to outweigh its benefits.

18: A: Something *banal* lacks originality and may be boring and trite. For example, a banal compliment is likely to be a common platitude. Like something that is inane, a banal compliment might be meaningless and lack a convincing quality or significance.

19. E: *Contrite* means to feel or express remorse, or to be regretful and interested in repenting. The noun *contrition* refers to severe remorse or penitence.

20. A: *Mollify* means to sooth, pacify, or appease. It usually is used to refer to reducing the anger or softening the feelings or temper of another person, or otherwise calm them down. For example, a customer service associate may need to mollify an irate customer who is furious about the defect in his or her purchase.

21. D: Of the provided choices, *fickle* is closest in meaning to *capricious.* A person who is capricious tends to display erratic or unpredictable behavior, which is similar to fickle, which is also likely to change spontaneously or behave erratically.

22. C: Although *paltry* often is used as an adjective to describe a very small or meager amount (of money, in particular), it can also mean something trivial or insignificant.

23. B: The word *shirk* means to evade and is often used in the context of shirking a responsibility, duty, or work.

24. E: *Assuage* most nearly means to soothe or comfort, as in to assuage one's fears. It can also mean to lessen or make less severe, or to relieve. For example, an ice pack on a swollen knee may assuage the pain.

25. A: Something that is *tacit* is usually unspoken but implied. Tacit approval, for example, occurs when agreement or approval is understood without being explicitly stated.

26. D: As a verb, *advocate* means to *support* something or someone.

27. A: To *corroborate* is a verb that means to verify or give support to.

28. E: *Sheer* is an adjective that means see-through or thin. The word most closely related to *sheer* is *translucent*. If something is *sheer* or *translucent*, you are able to see through it.

29. E: A *vow* is a solemn pledge, oath, or promise made to someone.

30. B: The word *incite* means to encourage or stir up disruptive behavior. A synonym for the word *incite* is *provoke*, which means to stimulate or stir up emotion in someone.

Analogies

1. C: This is a part/whole analogy. A chapter is a section, or portion, of a book. A paragraph is a section of an essay in much the same way that a chapter is in a book. The word pairs in Choices *A*, *B*, and *E* are best described as near synonyms, but not necessarily parts of one another. Choice *D*, *dialogue is to play*, does include more of a part-to-whole relationship, but dialogue is the way the story is conveyed in a play (like sentences in a book). A better matching analogy would be *scene* or *act is to play*, since plays are divided into scenes or larger acts.

2. A: This is a type of category analogy. A dress is a type of garment. Garment is the broad category, and dress is the specific example used. Diesel is a type of fuel, so it holds the same relationship. Month is not a type of year; it's part of the year. A suit is not a type of tie, but it might be worn with a tie. Clothing is not a type of wardrobe; it is stored in a wardrobe. Lastly, a coat isn't a type of winter; it is a garment worn in the winter.

3. E: This is a provider/provision analogy. The analogy focuses on where the mode of transportation is stored or housed when not in use. A garage is where a car is kept when not in use, much like a hanger for an airplane. Choice *C*, airport, might be an appealing choice, but an airport doesn't have as precise of a relationship to a plane as does a garage to a car. An airport might be where planes are located before use or where one might see a lot of planes, like a parking lot for a car.

4. C: This is an intensity analogy; *trickle* indicates low intensity while *gush* indicates high intensity. Fluid that is trickling is barely moving or of low volume, while gushing fluid is moving fast and often in a larger volume. The only choices that are related by intensity are Choices *C* and *D*, but *D* reverses the order within the relationship. Sob is a more significant cry versus a small sniffle. Tepid is lukewarm or slightly warm, while scorching is very hot.

5. B: This is a provider/provision analogy. The connection is between a condition and the medical professional who treats that condition. Acne may be treated by a dermatologist, a skin doctor. A cataract, an eye condition, is treated by an ophthalmologist. A psychologist treats psychological or mental conditions. An otolaryngologist is an ear, nose, and throat doctor. An otologist is an ear doctor, and an orthopedist treats skeletal issues.

6. C: This is a cause and effect analogy. The common thread is a physical reaction (effect) to a cause. Someone might jump from a surprise. Choice *C* maintains this relationship because someone might chuckle in response to a joke. Choice *E* does contain a type of physical reaction effect (crying) and a cause (sadness), but the order is reversed. Moreover, "sadness" is an emotion, whereas a surprise or a joke as opposed to an event or occurrence the same way that a surprise or joke is.

7. B: This is tool/use analogy. The connection is between a tool or utensil and the simple action that the tool is used for. This question is made easier by providing the next utensil, a fork. While a knife is used to slice things, a fork can be used to spear pieces of food so the pieces can be lifted to the mouth for eating. Therefore, Choice *B* is the best option. A fork is not used for cutting; that is the function of a knife. Therefore, Choice *A* is incorrect. Choices *C* and *D* are incorrect because they are not actions that a fork is used for. Choice *E, eat,* is technically an action the fork is used for, although it is not as precise as *spear,* making Choice *B* a better answer.

8. D: This is a grammatical analogy. It is comparing a verb in the simple present tense with the same verb in the simple past tense. *Ate* is the past tense of *eat,* and *spun* is the past tense of *spin.*

9. D: This is a synonyms analogy, which matches terms based on their similar meanings. *Obviate* is a verb that can mean to prevent or avoid, or to render something unnecessary. It is usually used with an object, one that is often a difficulty or disadvantage. For example, wearing a helmet while cycling can obviate the risk of a skull injury should a fall occur. *Preclude* is also a verb. It means to prevent something from occurring or existing. For example, a thunderstorm may preclude a picnic. It can also mean to exclude from something. For example, an inability to get wet after surgery would preclude the patient from swimming. Of the answer choices provided, Choice *D, appease is to placate,* is the only other pair of synonyms. Like the terms in the prompt, these words are both verbs. *Placate* means to pacify, calm, or make a person less angry or upset. Similarly, *appease* means to pacify or calm someone by fulfilling their demands.

10. A: This can be considered a type of characteristic analogy because it is matching a unit of measure with an example of what that unit measures. An *acre* is an example of a unit of measure for area. Plots of land can be measured in acres, and this value gives information about how much area of land that plot occupies. A *fathom* is an example of a unit to measure depth. It is typically used to refer to water depth.

11. A: This is a characteristic analogy. The connection lies in what color results when yellow is added to the primary color. Green is the secondary color that is formed when yellow is added to blue, just as orange is the secondary color created when yellow is added to red.

12. D: This is a parts/whole analogy. The connection is the outer removable, inedible layer of the fruit. The best choice is *D, shell is to coconut,* because the shell is the outer, inedible part of the coconut. While peaches do have fuzz, the fuzz is an edible portion of the skin. Therefore, Choice *E* is not as closely related to the question stem as the pair in Choice *D.*

13. E: This is a provider/provision analogy. The connection is a type of person or quality of a person and what that personality trait provides (the output, as a noun). A *sycophant* is someone who dishes out a lot of flattery, or insincere praise, often to better his or her situation. A *raconteur* dishes out or regales people with anecdotes and stories in an interesting way.

14. E: This is an antonyms analogy that matches adjectives with opposite meanings. A *viscous* fluid is thick and slowly moving, while a runny one is thin and flows freely. The only answer choice that is also a pair of adjectives that are antonyms is Choice *E.* When used as an adjective, *obscure* refers to something inconspicuous, unnoticeable, or ambiguous. It usually is used to refer to something with an unclear meaning or hard to understand, such as obscure intentions or the use of obscure language. *Unequivocal* is an adjective that means essentially the exact opposite: unambiguous or leaving no doubt.

15. B: This is another antonyms analogy. *Zenith* and *nadir* are astronomical nouns with opposite meanings that have also been adopted into conversational (non-technical) English. In an astronomical sense, the *zenith* is the point in the celestial sphere or sky that lies directly above the observer, while the *nadir* is the point directly below the observer. These terms have been incorporated into common language to mean the very top or culminating point of something (the *zenith*), and the very bottom, lowest, or worst (the *nadir*). In this question, the term *valley* is provided for the next pair. Like *nadir,* a *valley* is a very low point. The correct answer will then be a high point, which is best captured by Choice *B, pinnacle.*

16. E: This analogy matches a characteristic or type of person with the quality that type of person possesses. A *pundit* is an expert in a certain field or particular subject. A pundit has *expertise.* A scholar has learned or gained knowledge in the areas he or she has studied. A scholar has *erudition.*

17. E: This analogy matches an occupation with what someone in that job creates and uses as a plan for their work. An architect creates a blueprint to be a rendering of the plan for the structures that builders will use to erect the building, much like a composer creates a score that musicians will follow to play the music. The other answer choices do not maintain this same relationship.

18. C: This is a type of tool/user analogy. Rather than simply being a "tool" used by the user, it pairs the user with what they seek. A detective searches for evidence or clues, while a prospector searches for gold.

19. B: This is a type of tool/use analogy. It matches a tool and what it is used to measure. An odometer is used to measure distance traveled in a car. A caliper is tool used to measure thickness. For example, personal trainers use skinfold calipers to measure the thickness of various folds of skin to estimate body fat.

20. A: This is a category analogy. The connection is between a type of food and how that food is classified. A radish is a type of vegetable and a garbanzo bean (chickpea) is a type of legume. Pineapples are not a type of berry, lettuce is not a type of spinach, and cucumber is not a type of salad. Therefore, Choices *B, C,* and *D* are incorrect. Choice *E* may look appealing because citrus is a type of fruit, but "citrus" isn't a specific fruit. For the analogy to hold, it would need to be a specific citrus fruit, like lemon.

21. D: This is an intensity analogy. *Adore* is a stronger version of *appreciate. Loathe* is a stronger version of *dislike. Detest* and *hate,* Choices *A* and *B,* are more synonymous with *loathe* (rather than being different intensities), so they are not the best choices. Choice *C* is an antonym, and Choice *E* is unrelated.

22. E: This analogy makes use of the temporal relationship between an event (a holiday) and the calendar month that it occurs. Thanksgiving is a holiday in November, like Christmas is a holiday in December. Choice *C* is incorrect because Labor Day is in September. The other choices do not relate a specific holiday to the particular month in which they occur.

23. A: This is a category analogy. The common thread is the classification of the animal. An alligator is a type of reptile, just as an elephant is a type of mammal.

24. B: This is a source/comprised of analogy that focuses on pairing a raw material with an item that's created from it. Nylon is used to make parachutes just as neoprene is used to make wetsuits. Although sweaters are made from wool, the relationship is reversed. Test takers should remember to be careful about maintaining the same order of the words in the relationship when solving analogies; otherwise, the meaning is changed.

25. E: This is a tool/user analogy. The connection is the type of tool the artist uses to hold and create their work. An easel holds the paper or canvas that a painter uses to create a painting much like a loom is the apparatus used to hold and weave yarns into a blanket or other tapestry.

26. C: This is a synonyms analogy, which matches terms based on similar meanings. *Coarse* and *rough* mean the same thing, so they are synonyms. Choice *C* is the best answer choice available. *Funny* and *amusing* are synonyms, because they both mean that someone is acting comical. The other answer choices do not share the same meaning with each other.

27. E: This is an antonyms analogy, which matches terms that have opposite meanings. *Fluctuate* means to sway, waver, or change, so the word *persist*, which means to stay or remain stable, is the opposite of *fluctuate*. The word pair in the answer choices that have opposite meanings are *happy* and *sad*, Choice *E*.

28. B: This is a part to whole analogy. An *engine* is part of a car, just like a *branch* is part of a *tree*.

29. D: This analogy relates an object to its function. *Lotion* is an object used to moisturize or *hydrate*. Let's look at another object that shows its use through the answer choices. A *pot* is an object used to *boil*, so Choice *D* is the best answer.

30. D: This is a symbol/representation analogy. *Green* represents *go*. When people see a green light, especially driving on the road, it is a signal for them to keep driving. Similarly, a *dove* represents *peace*. In literature and history, doves often represent peace or calm.

Reading Comprehension

1. A: To show the audience one of the effects of criminal rehabilitation by comparison. Choice *B* is incorrect because although it is obvious the author favors rehabilitation, the author never asks for donations from the audience. Choices *C* and *D* are also incorrect. We can infer from the passage that American prisons are probably harsher than Norwegian prisons. However, the best answer that captures the author's purpose is Choice *A*, because we see an effect by the author (recidivism rate of each country) comparing Norwegian and American prisons.

2. C: A sense of foreboding. The narrator, after feeling excitement for the morning, feels "that something awful was about to happen," which is considered foreboding. The narrator mentions larks and weather in the passage, but there is no proof of anger or confusion at either of them.

3. B: To convince the audience that judges holding their positions based on good behavior is a practical way to avoid corruption. Choice A is incorrect because although he mentions the condition of good behavior as a barrier to despotism, he does not discuss it as a practice in the States. Choice C is incorrect because the author does not argue that the audience should vote based on judges' behavior, but rather that good behavior should be the condition for holding their office. Choice D is not represented in the passage, so it is incorrect.

4. B: Whatever happened in his life before he had a certain internal change is irrelevant. Choices *A, C,* and *D* use some of the same language as the original passage, like "revolution," "speak," and "details," but they do not capture the meaning of the statement. The statement is saying the details of his previous life are not going to be talked about—that he had some kind of epiphany, and moving forward in his life is what the narrator cares about.

5. C: A certain characteristic or quality imposed upon something else. The sentence states that "the impress of his dear hand [has] been stamped everywhere," regarding the quality of his tastes and creations on the house. Choice *A* is one definition of *impress*, but this definition is used more as a verb than a noun: "She impressed us as a songwriter." Choice *B* is incorrect because it is also used as a verb: "He impressed the need for something to be done." Choice *D* is incorrect because it is part of a physical act: "the businessman impressed his mark upon the envelope." The phrase in the passage is meant as figurative, since the workmen did most of the physical labor, not the Prince.

6. B: Mr. Ford assumed Booth's movement throughout the theater was due to being familiar with the theater. Choice *A* is incorrect; although Booth does eventually make his way to Lincoln's box, Mr. Ford does not make this distinction in this part of the passage. Choice *C* is incorrect; although the passage mentions "companions," it mentions Lincoln's companions rather than Booth's companions. Finally, Choice *D* is incorrect; the passage mentions "dress circle," which means the first level of the theater, but this is different from a "dressing room."

7. C: The tone of this passage is neutral since it is written in an academic/informative voice. It is important to look at the author's word choice to determine what the tone of a passage is. We have no indication that the author is excited, angry, or sorrowful at the effects of bacteria on milk, so Choices *A, B,* and *D* are incorrect.

8. D: A city whose population is made up of people who seek quick fortunes rather than building a solid business foundation. Choice *A* is a characteristic of Portland, but not that of a boom city. Choice *B* is close—a boom city is one that becomes quickly populated, but it is not necessarily always populated by residents from the east coast. Choice *C* is incorrect because a boom city is not one that catches fire frequently, but one made up of people who are looking to make quick fortunes from the resources provided on the land.

9. B: A myth. Some key words indicating the type of passage are "deities," "Greek hero," "centaur," and the name of the goddess Artemis. A eulogy is typically a speech given at a funeral, making Choice *A* incorrect. Choices *C* and *D* are incorrect, as "virgin huntresses" and "centaurs" are typically not found in historical or professional documents.

10. B: Temporary resident. Although we don't have much context to go off of, we know that one is probably not a "lifetime partner" or "farm crop" in civilized life. These two do not make sense, so Choices *C* and *D* are incorrect. Choice *A* is also a bit strange. To be an "illegal immigrant" in civilized life is not a used phrase, making it incorrect.

11. B: To retain the truth of the work. The author says that "music and rhythm and harmony are indeed fine things, but truth is finer still," which means that the author stuck to a literal translation instead of changing up any words that might make the English language translation sound better.

12. C: Choice *C* is correct because the word *unremarkable* should be changed to *remarkable* to be consistent with the details of the passage. This question requires close attention to the passage. Choice *A* is incorrect; it can be found where the passage says, "no less than six named and several unnamed varieties of the peach have thus produced several varieties of nectarine." Choice *B* is incorrect; it can be found where the passage says, "it is highly improbable that all these peach-trees . . . are hybrids from the peach and nectarine." Choice *D* is incorrect because we see in the passage that "the production of peaches from nectarines, either by seeds or buds, may perhaps be considered as a case of reversion."

13. B: Choice *B* is correct because the meaning holds true even if the words have been switched out or rearranged some. Choice *A* is incorrect because it has trees either bearing peaches or nectarines, and the trees in the original phrase bear both. Choice *C* is incorrect because the statement does not say these trees are "indifferent to bud-variation," but that they have "indifferently [bore] peaches or nectarines." Choice *D* is incorrect; the statement may use some of the same words, but the meaning is skewed in this sentence.

14. C: Choice *C* is correct; we cannot infer that the passage takes place during the nighttime. While we do have a statement that says that the darkness thickened, this is the only evidence we have. The darkness could be thickening because it is foggy outside. We don't have enough proof to infer otherwise. Choice *A* is incorrect; some of the evidence here is that "the cold became intense," and people were decorating their shops with "holly sprigs," a Christmas tradition. It also mentions that it is Christmas time at the end of the passage. Choice *B* is incorrect; we *can* infer that the narrative is located in a bustling city street by the actions in the story. People are running around trying to sell things, the atmosphere is busy, there is a church tolling the hours, etc. The scene switches to the Mayor's house at the end of the passage, but the answer says *majority,* so this is still incorrect. Choice *D* is incorrect; we *can* infer that the Lord Mayor is wealthy—he lives in the "Mansion House" and has fifty cooks.

15. A: Choice *A* is correct. *Ruddy* means "red," so we can deduce that the phrase *made pale faces ruddy* means that the shops made people's faces look red. This is a descriptive sentence, so a careful reading of what's going on is imperative. Choices *B, C,* and *D,* although they may contain components of the original meaning, are incorrect.

16. B: Choice *B* is correct. We can use context clues to solve this. The passage states, "if you are a novice and accustomed to walking only over smooth and level ground," which implies that whoever a "novice" is would only be accustomed to smooth ground, not rough ground. An expert would be accustomed to rockier ground than the passage indicates.

17. B: Choice *B* is correct because the sentence states, "In unknown regions take a responsible guide with you, unless the trail is short, easily followed, and a frequented one." Choice *A* is incorrect; the passage does not state that you should try and explore unknown regions. Choice *C* is incorrect; the passage talks about trails that contain pitfalls, traps, and boggy places, but it does not say that *all* unknown regions contain these things. Choice *D* is incorrect; the passage mentions "rail" and "boat" as means of transport at the beginning, but it does not suggest it is better to travel unknown regions by rail.

18. D: Choice *D* is correct because, although it may be real advice an experienced hiker would give to an inexperienced hiker, the question asks about details in the passage, and this is not in the passage. Choice *A* is incorrect; we do see the author encouraging the reader to learn about the trail beforehand: "wet or dry; where it leads; and its length." Choice *B* is also incorrect; we do see the author telling us the time will lengthen with boggy or rugged places opposed to smooth places. Choice *C* is incorrect; at the end of the passage, the author tells us, "do not go alone through lonely places . . . unless you are quite familiar with the country and the ways of the wild."

19. D: Choice *D* is correct because it is not a detail in the passage, although it is close to a detail in the passage. The passage says that the speaker threw the door open as a child would when they expected to see a spectre, but this doesn't mean that the speaker sees a spectre himself. Choices *A, B,* and *C* are incorrect because they can be found as details in the passage.

20. B: Choice *B* is correct. The passage details the speaker and his friend going to the apartment. The speaker, terrified of finding the "creature," realizes that the creature is gone after he barges in through the door. Choice *A* is incorrect because we never see the speaker and the friend trying to hunt anything. Choice *C* is incorrect because we do not have proof that the friend let loose a terrifying creature in the speaker's house. Choice *D* is also incorrect because we have no proof that the speaker is "eager" to show the creature. We see that the speaker is more reluctant to find the creature inside the apartment than eager and does not wish to show any of the friends.

Clerical

1. D: The word *flippant* is a synonym for the word *facetious.*

2. A: The word *poignent* should be spelled *poignant.* The words *reciprocity, bourgeois,* and *ramification* are all spelled correctly.

3. B: Choice *B*, 21 18 15 12, is the correct answer. The pattern is to subtract three from each number.

4. C: The correct answer is 12, Choice *C.* Solve the problems in parentheses first: $(6 - 2) = 4$ and $(9 \div 3) = 3$. Then, multiply those results $(4 x 3)$ to get 12.

5. B: Choice *B*, XRD7903518&, is the exact copy of the code.

6. A: The correct answer is Choice *A, deteriorating. Relapse* means setback, *digress* means wander, and *imminent* means looming.

7. A: The correct answer is Choice *A, agreement. Clamor, tumult,* and *dissonance* are all synonyms for *discord.*

8. D: The correct answer is Choice *D*. 770 x 139 = 107,030.

9. A: The correct answer is Choice *A*: 7 x 7 x 7 x 7 x 7 = 16,807

10. B: The correct answer is Choice *B*, 10.5. Steven ate 2 apples, leaving him with 6. Helen gave him 4.5 apples: $6 + 4.5 = 10.5$.

11. D: The correct answer is D, *careful*. The word *trite* means stale, *sloppy* means messy, and *stunted* means underdeveloped.

12. A: The correct answer is Choice *A*, *receptive*. *Aggressive*, *inimical*, and *intimidating* are all synonyms for hostile.

13. C: The correct answer is Choice *C*, $515. First, you would multiply $\$11 \times 25 = \$275$, and then multiply $\$15 \times 16 = \$240$. Then, add the two amounts up to get $515:

$$\$275 + \$240 = \$515$$

14. C: The correct answer is Choice *C*, 23791 Maple Court Dr.

15. B: The correct answer is Choice *B*, *verbose*. *Concise*, *brief*, and *pithy* are all synonyms for *succinct*.

16. B: The correct answer is Choice *B*, second. The correct order of names is Amber Greene, Avery Greenly, Ashley Gregson, and Aaron Greily.

17. C: The correct answer is Choice *C*, an Asian-American man who is 5'11."

18. D: The correct answer is Choice *D*, C-M-6.

19. D: The correct answer is Choice *D*, 4, 3, 1, 2, 5. These are the numbers that go in descending order: 11.1, 10.5, 9 ¾, 9 ½, and $8\ ^5/_8$.

20. A: The correct answer is Choice *A*, *perceptive*. *Ignorant* means unaware, *unknowing* means naïve, and *blatant* means obvious.

21. B: The correct answer is Choice *B*. $397\ x\ 863 = 342{,}611$.

22. C: The correct answer is Choice *C*, HRC7816$3L0.

23. B: The correct answer is Choice *B*, *cartharsys*, which should be spelled *cartharsis*. *Prurient*, *visionary*, and *treble* are all spelled correctly.

24. A: The correct answer is Choice *A*, 98. Each number on the bottom is divided by 343 to get the number above it. Dividing 33,614 by 343 equals 98.

25. D: The correct answer is Choice *D*, 288. The pattern is that each number is multiplied by 2. 144 multiplied by 2 is 288.

26. C: The correct answer is Choice *C*, 1.75%, because $35 \div 2000 = 0.0175$.

27. D: The correct answer is Choice *D*, *benevolent*, which means kind. *Wicked*, *malicious*, and *malevolent* are all synonyms for malignant.

28. C: The correct answer is Choice *C, separately. Arbitrarily* means randomly, *inordinately* means slightly, and *incommensurately* means inadequately.

29. A: The correct answer is Choice *A*, ^**#DL7634IIX.

30. B: The correct answer is Choice *B*, 16.04.

31. A: The correct answer is Choice *A, bovine. Definat* should be spelled *definite, superbe* should be *superb*, and *thorought* should be *throughout.*

32. C: The correct answer is Choice *C, emeritus. Present, incorporated,* and *unreserved* are all antonyms for *retired.*

33. B: The correct answer is Choice *B*: $350 \, x \, 79 = 27{,}650$.

34. C: The correct answer is Choice *C*: Lucas, Louis, Linwood, Lewis, Leonard.

35. D: The correct answer is Choice *D*: $815 + 369 = 1{,}184$.

36. A: The correct answer is Choice *A*, 9 hours. There are three nights left with three hours of study time each:

$$3 \times 3 = 9$$

37. B: The correct answer is Choice *B*, two. Los Angeles and Kansas City have populations higher than 450,000.

38. A: The correct answer is Choice *A*. To find the average, add 179,883 and 389,938 and then divide by 2.

39. A: The correct answer is Choice *A*, Missouri.

40. C: The correct answer is Choice *C*, 169. To solve 13^2, multiply $13 \, x \, 13$.

41. B: The correct answer is Choice *B*, 64. Each number on the bottom is multiplied by 8 to get the number on the top. 8 multiplied by 8 is 64.

42. D: The correct answer is Choice *D, skirt. Confront* means challenge, *consume* means ingest, and *validate* means certify.

43. C: The correct answer is Choice *C, pardon. Complaint, charge,* and *warrant* are all synonyms for *indictment.*

44. B: The correct answer is Choice *B*: While *you're* out, pick up *your* clothes from the dry cleaner. *You're* is a contraction of *you are*, and *your* is possessive.

45. C, D: The correct answers are Choices *C* and *D*. As written, the sentence has two standalone phrases separated by a comma splice. Choice *C* uses a conjunction to join the two phrases. Choice *D* separates the phrases into two sentences.

46. A: The correct answer is Choice *A*, 19,683. 3^9 is 3 multiplied by 3 nine times, or $3 \, x \, 3 \, x \, 3 \, x \, 3 \, x \, 3 \, x \, 3 \, x \, 3 \, x \, 3 \, x \, 3$.

47. B: The word *unfavorable* is a synonym for *adverse.*

48. C: The correct answer is Choice *C*: Regan, Richard, Rick, Riley, Ryan.

49. A: The correct answer is Choice *A*: $1683 - 765 = 918$.

50. D: The correct answer is Choice *D*, ALSX13390V.

Mathematics

1. B: The fraction $\frac{12}{60}$ can be reduced to $\frac{1}{5}$, which puts the fraction in lowest terms. First, it must be converted to a decimal. Dividing 1 by 5 results in 0.2. Then, to convert to a percentage, move the decimal point two units to the right and add the percentage symbol. The result is 20%.

2. B: Common denominators must be used. The LCD is 15, and $\frac{2}{5} = \frac{6}{15}$. Therefore,

$$\frac{14}{15} + \frac{6}{15} = \frac{20}{15}$$

In lowest terms, the answer is $\frac{4}{3}$. A common factor of 5 was divided out of both the numerator and denominator.

3. A: A product is found by multiplying. Multiplying two fractions together is easier when common factors are cancelled first to avoid working with larger numbers.

$$\frac{5}{14} \times \frac{7}{20}$$

$$\frac{5}{2 \times 7} \times \frac{7}{5 \times 4}$$

$$\frac{1}{2} \times \frac{1}{4} = \frac{1}{8}$$

4. D: Division is completed by multiplying by the reciprocal. Therefore:

$$24 \div \frac{8}{5} = \frac{24}{1} \times \frac{5}{8}$$

$$\frac{3 \times 8}{1} \times \frac{5}{8}$$

$$\frac{15}{1} = 15$$

5. C: Common denominators must be used. The LCD is 168, so each fraction must be converted to have 168 as the denominator.

$$\frac{5}{24} - \frac{5}{14} = \frac{5}{24} \times \frac{7}{7} - \frac{5}{14} \times \frac{12}{12}$$

$$\frac{35}{168} - \frac{60}{168} = -\frac{25}{168}$$

6. C: The correct mathematical statement is the one in which the smaller of the two numbers is on the "less than" side of the inequality symbol. It is written in answer *C* that $\frac{1}{3} > -\frac{4}{3}$, which is the same as $-\frac{4}{3} < \frac{1}{3}$, a correct statement.

7. C: $-\frac{1}{5} > \frac{4}{5}$ is incorrect. The expression on the left is negative, which means that it is smaller than the positive expression on the right. As it is written, the inequality states that the expression on the left is greater than the expression on the right, which is not true.

8. D: This is a one-step, real-world application problem. The unknown quantity is the number of cases of cola to be purchased. Let x be equal to this amount. Because each case costs $3.50, the total number of cases multiplied by $3.50 must equal $40. This translates to the mathematical equation $3.5x = 40$. Divide both sides by 3.5 to obtain $x = 11.4286$, which has been rounded to four decimal places. Cases are sold whole (the store does not sell portions of cases), and there is not enough money to purchase 12 cases. Therefore, there is only enough money to purchase 11 cases.

9. A: Rounding can be used to find the best approximation. All of the values can be rounded to the nearest thousand. 15,412 SUVs can be rounded to 15,000. 25,815 station wagons can be rounded to 26,000. 50,412 sedans can be rounded to 50,000. 8,123 trucks can be rounded to 8,000. Finally, 18,312 hybrids can be rounded to 18,000. The sum of the rounded values is 117,000, which is closest to 120,000.

10. D: There are 52 weeks in a year, and if the family spends $105 each week, that amount is close to $100. A good approximation is $100 a week for 50 weeks, which is found through the product $50 \times 100 = \$5{,}000$.

11. D: There were 48 total bags of apples sold. If 9 bags were Granny Smith and the rest were Red Delicious, then 48 – 9 = 39 bags were Red Delicious. Therefore, the ratio of Granny Smith to Red Delicious is 9:39.

12. B: The average rate of change is found by calculating the difference in dollars over the elapsed time. Therefore, the rate of change is equal to ($4,900–$4,000)÷3 months, which is equal to $900÷3, or $300 a month.

13. A: Let x be the unknown, the number of hours Erin can work. We know Katie works $2x$, and the sum of all hours is less than 21. Therefore,

$$x + 2x < 21$$

Which simplifies into $3x < 21$. Solving this results in the inequality $x < 7$ after dividing both sides by 3. Therefore, Erin worked less than 7 hours.

14. B: If a calculator is used, divide 33 into 14 and keep two decimal places. If a calculator is not used, multiply both the numerator and denominator times 3. This results in the fraction $\frac{42}{99}$, and hence a decimal of 0.42.

15. C: Gina answered 60% of 35 questions correctly; 60% can be expressed as the decimal 0.60. Therefore, she answered $0.60 \times 35 = 21$ questions correctly.

16. B: The unknown quantity is the number of total questions on the test. Let x be equal to this unknown quantity. Therefore, $0.75x = 12$. Divide both sides by 0.75 to obtain $x = 16$.

17. A: First, use the distributive property on the left side. This results in:

$$3x + 6 = 14x - 5$$

Then use the addition property to add 5 to both sides, and then to subtract $3x$ from both sides, resulting in $11 = 11x$. Finally, use the multiplication property to divide each side by 11. Therefore, $x = 1$ is the solution.

18. C: First, collect like terms to obtain:

$$12 - 5x = -5x + 12$$

Then, use the addition principle to add $5x$ to both sides, resulting in the mathematical statement $12 = 12$. This is always true; therefore, all real numbers satisfy the original equation.

19. D: Use the distributive property on both sides to obtain:

$$4x + 20 + 6 = 4x + 6$$

Then, collect like terms on the left, resulting in:

$$4x + 26 = 4x + 6$$

Next, use the addition principle to subtract $4x$ from both sides, which results in the false statement $26 = 6$. Therefore, there is no solution.

20. A: First, define the variables. Let x be the first integer; therefore, $x + 1$ is the second integer. This is a two-step problem. The sum of three times the first and two less than the second is translated into the following expression:

$$3x + (x + 1 - 2)$$

Set this expression equal to 411:

$$3x + (x + 1 - 2) = 411$$

Simplify the left-hand side is to obtain:

$$4x - 1 = 411$$

Use the addition and multiplication properties to solve for x. First, add 1 to both sides and then divide both sides by 4 to obtain $x = 103$. The next consecutive integer is 104.

21. B: First, the information is translated into the ratio $\frac{15}{80}$. To find the percentage, translate this fraction into a decimal by dividing 15 by 80. The corresponding decimal is 0.1875. Move the decimal point two places to the right to obtain the percentage 18.75%.

22. B: If sales tax is 7.25%, the price of the car must be multiplied by 1.0725 to account for the additional sales tax. Therefore,

$$15{,}395 \times 1.0725 = 16{,}511.1375$$

This amount is rounded to the nearest cent, which is $16,511.14.

23. A: The bar chart shows how many men and women prefer each genre of movies. The dark gray bars represent the number of women, while the light gray bars represent the number of men. The light gray bars are higher and represent more men than women for the genres of Comedy and Action.

24. B: A line graph represents continuous change over time. A bar graph may show change but not necessarily continuous change over time. A pie graph is better for representing percentages of a whole. Histograms are best used in grouping sets of data in bins to show the frequency of a certain variable.

25. C: The mean for the number of visitors during the first 4 hours is 14. The mean is found by calculating the average for the four hours. Adding up the total number of visitors during those hours gives:

$$12 + 10 + 18 + 16 = 56$$

Then:

$$56 \div 4 = 14$$

26. C: The mode for a set of data is the value that occurs the most. The grade that appears the most is 95. It is the only value that repeats in the set.

27. B: The relationship between age and attention span is a positive correlation because the general trend for the data is up and to the right. As age increases, so does attention span.

28. C: There are 0.006 kiloliters in 6 liters because 1 liter=0.001kiloliters. The conversion comes from the chart where the prefix kilo is found three places to the left of the base unit.

29. B: The only relation in which every x-value corresponds to exactly one y-value is the relation given in Choice *B*, making it a function. The other relations have the same first component paired up to different second components, which goes against the definition of a function.

30. A: To find a function value, plug in the number given for the variable and evaluate the expression, using the order of operations (parentheses, exponents, multiplication, division, addition, subtraction). The function given is a polynomial function:

$$f(5) = 5^2 - 2 \times 5 + 1$$
$$25 - 10 + 1 = 16$$

31. C: The function given is a polynomial function. Anything can be plugged into a polynomial function to get an output. Therefore, its domain is all real numbers, which is expressed in interval notation as $(-\infty, \infty)$.

32. D: The smallest number in this function's range occurs when plugging 0 into the function $f(0) = 5$. Any other output is a number larger than 5, even when a positive number is plugged in. When a negative number gets plugged into the function, the output is positive, and same with a positive number. Therefore, the domain is written as $[5, \infty)$ in interval notation.

33. C: First set the functions equal to one another, resulting in:

$$x^2 + 3x + 2 = 4x + 4$$

This is a quadratic equation, so the best way to find the answer is to convert it to standard form by subtracting $4x + 4$ from both sides of the equation, setting it equal to 0:

$$x^2 - x - 2 = 0$$

Then factor the equation:

$$(x - 2)(x + 1) = 0$$

Setting both factors equal to zero results in $x = 2$ and $x = -1$.

34. D: There will be no more coyotes when the population is 0, so set y equal to 0 and solve the quadratic equation:

$$0 = -(x - 2)^2 + 1600$$

Subtract 1600 from both sides and divide by -1. This results in:

$$1600 = (x - 2)^2$$

Then, take the square root of both sides. This process results in the following equation:

$$\pm 40 = x - 2$$

Adding 2 to both sides results in two solutions: $x = 42$ and $x = -38$. Because the problem involves years after 2000, the only solution that makes sense is 42. Add 42 to 2000; therefore, in 2042 there will be no more coyotes.

35. B: The ball is back at the starting point when the function is equal to 800 feet. Therefore, this results in solving the equation:

$$800 = -32t^2 + 90t + 800$$

Subtract 800 off of both sides and factor the remaining terms to obtain:

$$0 = 2t(-16t + 45)$$

Setting both factors equal to 0 result in $t = 0$, which is when the ball was thrown up initially, and:

$$t = \frac{45}{16} = 2.8 \text{ seconds}$$

Therefore, it will take the ball 2.8 seconds to come back down to its staring point.

36. B: Given a rational function, the expression in the denominator can never be equal to 0. To find the domain, set the denominator equal to 0 and solve for *x*. This results in $2 - x = 0$, and its solution is $x = 2$. This value needs to be excluded from the set of all real numbers, and therefore the domain written in interval notation is:

$$(-\infty, 2) \cup (2, \infty)$$

37. C: The Pythagorean Theorem can be used to find the missing length *x* because it is a right triangle. The theorem states that:

$$6^2 + 8^2 = x^2$$

Which simplifies into $100 = x^2$. Taking the positive square root of both sides results in the missing value $x = 10$.

38. A: The probability of .9 is closer to 1 than any of the other answers. The closer a probability is to 1, the greater the likelihood that the event will occur. The probability of 0.05 shows that it is very unlikely that an adult driver will wear their seatbelt because it is close to zero. A zero probability means that it will not occur. The probability of 0.25 is closer to zero than to one, so it shows that it is unlikely an adult will wear their seatbelt.

39. B: The goal is to first isolate the variable. The fractions can easily be cleared by multiplying the entire inequality by 5, resulting in:

$$35 - 4x < 3$$

Then, subtract 35 from both sides and divide by -4. This results in $x > 8$. Notice the inequality symbol has been flipped because both sides were divided by a negative number. The solution set, all real numbers greater than 8, is written in interval notation as $(8, \infty)$. A parenthesis shows that 8 is not included in the solution set.

40. D: This system can be solved using the method of substitution. Solving the first equation for *y* results in:

$$y = 14 - 2x$$

Plugging this into the second equation gives:

$$4x + 2(14 - 2x) = -28$$

Which simplifies to $28 = -28$, an untrue statement. Therefore, this system has no solution because no *x* value will satisfy the system.

41. B: The slopes of perpendicular lines are negative reciprocals, meaning their product is equal to -1. The slope of the line given needs to be found. Its equivalent form in slope-intercept form is:

$$y = -\frac{4}{7}x + \frac{23}{7}$$

So, its slope is $-\frac{4}{7}$. The negative reciprocal of this number is $\frac{7}{4}$. The only line in the options given with this same slope is:

$$y = \frac{7}{4}x - 12$$

42. C: This system can be solved using substitution. Plug the second equation in for *y* in the first equation to obtain:

$$2x - 8x = 6$$

Which simplifies to $-6x = 6$. Divide both sides by 6 to get $x = -1$, which is then back-substituted into either original equation to obtain $y = -8$.

43. A: Division can be used to solve this problem. The division necessary is:

$$\frac{5.972 \times 10^{24}}{7.348 \times 10^{22}}$$

To compute this division, divide the constants first then use algebraic laws of exponents to divide the exponential expression. This results in about 0.8127×10^2, which written in scientific notation is 8.127×10^1.

44. A: The formula for the rate of change is the same as slope: change in *y* over change in *x*. The *y*-value in this case is percentage of smokers and the *x*-value is year. The change in percentage of smokers from 2000 to 2015 was 8.1 percent. The change in *x* was:

$$2000 - 2015 = -15$$

Therefore:

$$^{8.1\%}/_{-15} = -0.54\%$$

The percentage of smokers decreased 0.54 percent each year.

45. D: Let *x* be the unknown number. The difference indicates subtraction, and sum represents addition. To triple the difference, it is multiplied by 3. The problem can be expressed as the following equation:

$$3(5 - x) = x + 5$$

Distributing the 3 results in:

$$15 - 3x = x + 5$$

Subtract 5 from both sides, add $3x$ to both sides, and then divide both sides by 4. This results in:

$$\frac{10}{4} = \frac{5}{2} = 2.5$$

46. A: A proportion should be used to solve this problem. The ratio of tagged to total deer in each instance is set equal, and the unknown quantity is a variable *x*. The proportion is:

$$\frac{300}{x} = \frac{5}{400}$$

Cross-multiplying gives $120{,}000 = 5x$, and dividing through by 5 results in 24,000.

47. B: First, the common factor 2 can be factored out of both terms, resulting in $2(y^3 - 64)$. The resulting binomial is a difference of cubes that can be factored using the rule:

$$a^3 - b^3 = (a - b)(a^2 + ab + b^2)$$

with a = y and b = 4. Therefore, the result is:

$$2(y - 4)(y^2 + 4y + 16)$$

48. D: The exponential rules:

$$(ab)^m = a^m b^m \text{ and } (a^m)^n = a^{mn}$$

They can be used to rewrite the expression as:

$$4^4 y^{12} \times 3^2 y^{14}$$

The coefficients are multiplied together and the exponential rule:

$$a^m a^n = a^{m+n}$$

Is then used to obtain the simplified form $2304y^{26}$.

49. B: The number of representatives varies directly with the population, so the equation necessary is $N = k \times P$, where N is number of representatives, k is the variation constant, and P is total population in millions. Plugging in the information for New York allows k to be solved for. This process gives $27 = k \times 20$, so $k = 1.35$. Therefore, the formula for number of representatives given total population in millions is:

$$N = 1.35 \times P$$

Plugging in $P = 11.6$ for Ohio results in $N = 15.66$, which rounds up to 16 total Representatives.

50. B: The outlier is 35. When a small outlier is removed from a data set, the mean and the median increase.

Greetings!

First, we would like to give a huge "thank you" for choosing us and this study guide for your Civil Service exam. We hope that it will lead you to success on this exam and for your years to come.

Our team has tried to make your preparations as thorough as possible by covering all of the topics you should be expected to know. In addition, our writers attempted to create practice questions identical to what you will see on the day of your actual test. We have also included many test-taking strategies to help you learn the material, maintain the knowledge, and take the test with confidence.

We strive for excellence in our products, and if you have any comments or concerns over the quality of something in this study guide, please send us an email so that we may improve.

As you continue forward in life, we would like to remain alongside you with other books and study guides in our library. We are continually producing and updating study guides in several different subjects. If you are looking for something in particular, all of our products are available on Amazon. You may also send us an email!

Sincerely,
APEX Test Prep
info@apexprep.com

Free Study Tips DVD

In addition to the tips and content in this guide, we have created a FREE DVD with helpful study tips to further assist your exam preparation. **This FREE Study Tips DVD provides you with top-notch tips to conquer your exam and reach your goals.**

Our simple request in exchange for the strategy-packed DVD is that you email us your feedback about our study guide. We would love to hear what you thought about the guide, and we welcome any and all feedback—positive, negative, or neutral. It is our #1 goal to provide you with top quality products and customer service.

To receive your **FREE Study Tips DVD**, email freedvd@apexprep.com. Please put "FREE DVD" in the subject line and put the following in the email:

a. The name of the study guide you purchased.

b. Your rating of the study guide on a scale of 1-5, with 5 being the highest score.

c. Any thoughts or feedback about your study guide.

d. Your first and last name and your mailing address, so we know where to send your free DVD!

Thank you!

Made in the USA
Monee, IL
04 October 2021